GEOPOLITICS OF OIL

GEOPOLITICS OF OIL

KATHERINE T. HARRIS
EDITOR

Nova Science Publishers, Inc.
New York

For permission to use material from this book please contact us:
Telephone 631-231-7269; Fax 631-231-8175
Web Site: http://www.novapublishers.com

NOTICE TO THE READER

LIBRARY OF CONGRESS CATALOGING-IN-PUBLICATION DATA

Harrisá, Katherine T.
Geopolitics of oil / Katherine T. Harrisá.
p. cm.
ISBN 978-1-60692-810-3 (hardcover)
1. Petroleum industry and trade--Government policy--United States. 2. Globalization. I. Title.
HD9560.6.H37 2009
338.2'72820973--dc22

2008055217

Published by Nova Science Publishers, Inc. ✝ New York

CONTENTS

PREFACE

The price and availability of oil and natural gas affects every American. That's why it is important that the Energy Committee's first hearing of the 110th Congress focused on the global oil situation and its implications for U.S. economic and national security interests. As a nation, we now depend on oil imports to meet sixty percent of our oil needs. Even modest disruptions in the world supply can result in price spikes at the pump, as we have seen in recent years. First, there is little surplus production capacity relative to global demand. Much of the current production is controlled by national oil companies that are often making political rather than economic decisions, and are not making the investments needed to maintain and expand production capacity.this book examines the geopolitical factors surrounding the acquisition of oil supplies.

Chapter 1

OPENING STATEMENT OF HON. JEFF BINGAMAN, U.S. SENATOR FROM NEW MEXICO*

The CHAIRMAN. Why don't we go ahead and get started. I thank you all very much for coming. The Senate still is in the process of getting its organizing resolution done, so Senator Domenici is officially the Chair, but he has allowed me to go ahead today, as we had intended when we first thought about this hearing.

Let me just point out the new members—and I know Senator Domenici probably wants to do the same—and welcome them. We have three new members on the Democratic side and three new members on the Republican side. Senators Lincoln and Sanders and Tester are our new members on the Democratic side. We welcome them. Senator Tester is here right now. On the Republican side, Senators DeMint, Corker and Sessions. And, of course, we welcome them as well.

Let me just briefly go through an opening statement here and then defer to Senator Domenici.

I think the idea of this hearing was to try to look at the big picture, begin the year with sort of an overview of the geopolitics of oil and I hope that that's a useful thing. There is a quote that my staff dug out of the files of the committee, from Scoop Jackson, when he chaired this committee back in 1980, and he said, at that time, ''The world will witness a growing struggle for secure access to oil through the end of this century and into the next. This gathering energy crisis deserves the highest priority in the counsels of Government. Few other problems

---

* Excerpted from S. HRG. 110–6, First Session to Receive Testimony on the Geopolitics of Oil and Its Implications for U.S. Economic and National Security, dated January 10, 2007.

are more complicated, few other problems will be more difficult to resolve. Moreover, many of the policies we are currently pursuing to deal with the energy crisis are only making it worse.'' So that was Senator Jackson's view of things in 1980.

I think, today, we still have the struggle for access to oil that he referred to. We also, of course, have a competition among consumers that has developed, particularly with the increasing appetite for oil in places like China and India. There are great implications for the United States in all of that, both for our economy and for our national security and the purpose of this hearing is to get some of the people who have thought about these issues to give us their views and then give us a chance to ask some questions.

So let me defer to Senator Domenici for any comments he has.

Chapter 2

Statement of Hon. Pete V. Domenici, U.S. Senator from New Mexico

Senator DOMENICI. Thank you very much, Senator Bingaman, and thanks to the witnesses for helping us today with their presentations that the Senators were interested in. I think it is most beneficial that we put into perspective who owns access to the oil in the world today. It is rather frightening when you get just that picture before you and nothing more, to know how things have changed dramatically and how little of the oil of the world is owned by the American companies that we are constantly arguing with and how little these oil companies of America have access and/or control over these oils.

I have had staff reduce the world's oil to a chart that shows where we are, and there is no question that private investors are already at a disadvantage. The rise in national oil companies has decreased access to reserves through the use of strategic energy agreements between governments. U.S. companies are being squeezed out.

Examples are the Chinese national oil company's development of an energy production agreement in Sudan and Iran, Russia's reclaiming of oil producing assets from Yukos to form a state oil company and just yesterday, Venezuelan President Hugo Chavez called for the end to foreign ownership of crude oil refineries in the Orinoco region. This activity further limits investment opportunities for investor-owned companies.

These trends are doubly concerning, given the many producer nations, political instability, and the lack of a legal system for enforcement of contract rights resulting in only a sufficient capital investment in the infrastructure necessary to sustain existing production, much less new capital on line.

For example, the recent prediction by one scholar of the extinction of the Iranian oil exports by 2014-2015, will they allow it to happen or will it be forestalled by investments from other countries that are less than friendly to U.S. interests?

I'd ask that the remainder of my remarks be made part of the record and thank you, Senator Bingaman, for opening the year with a bit of realism. Thank you. [The prepared statements of Senators Domenici, Smith and Sanders follow:]

PREPARED STATEMENT OF HON. PETE V. DOMENICI, U.S. SENATOR FROM NEW MEXICO

I would like to call this hearing to order.

No, you are not imagining things—this is the gavel in my hand.

First, I'd like to congratulate Senator Bingaman on being elected Chairman of the Committee. I look forward to continuing to work with him in the fine bipartisan tradition of this Committee.

Senator Bingaman has a commendable desire to get a running start on the Committee's work. When we scheduled this hearing, we didn't realize that he would not yet officially be Chairman.

So, for the time being, you get to call me ''Mr. Chairman''—even though the role is strictly ceremonial in this case.

I do appreciate Senator Bingaman planning this hearing on the Global Oil Balance and Its Implications for U.S. Economic and National Security. In a minute or two, I will hand the gavel over to him to preside. But first, I will share a few thoughts of my own.

As I have stated many times before, energy security is a complex issue that cannot be reduced to a ''sound bite.'' That being said, I will briefly summarize what I think are some important issues facing this nation.

I think it is useful that Chairman Bingaman is holding this hearing first. In this area, there is great risk in making decisions based on what we ''know'' to be true from the past—while the reality is that the world has changed.

Today we will hear about the dangers of our dependence on foreign sources of oil. But what may be even more frightening—and difficult to deal with—is the ways in which the very structure of oil markets have changed.

Most Americans, and many politicians, focus on large multinational corporations as the face of ''big oil.'' However, today we will hear that the reality is that National Oil Companies—those owned by foreign governments—control some three-quarters of the world's oil reserves.

Thus, today, we are largely dependent on supplies of oil controlled by governments whose values and priorities are often in conflict with America's.

Why do these changes in world oil markets matter with respect to our government's policy? Because approaches that focus on' the business practices of large private corporations will have little effect on world oil markets other than to disadvantage private investment by U.S. companies and harm our consumers.

Private investors are already at a disadvantage. The rise in National Oil Companies has decreased access to reserves through the use of strategic energy agreements between governments. U.S. companies are being squeezed out.

Examples are:

- The Chinese national oil companies' development of energy production agreements in Sudan and Iran; and
- Russia's reclaiming of oil producing assets from Yukos to form a new state oil company.
- Just yesterday,Venezuelan President Hugo Chavez called for the end to foreign ownership of crude oil refineries in the Orinoco region.

This activity further limits investment opportunities for investor-owned oil companies.

These trends are doubly concerning. In many producer nations, political instability and a lack of a legal system for the enforcement of contract rights results in insufficient capital investment in the infrastructure necessary to sustain existing production, much less bring new capacity on line.

Thus, for example, the recent prediction by one scholar of the ''extinction'' of Iranian oil exports by 2014-2015. Will this be allowed to happen? Or, will it be forestalled by investment from other states that are less than friendly to U.S. interests?

One option we don't have is to simply pretend that these trends will go away by themselves. We must take a two-pronged approach: First, we must do what we can to work toward U.S. energy security.

In 2005, we approved a comprehensive energy bill that is already showing results in many areas, including:

- the production and use of alternative fuels, and
- providing for a nuclear renaissance.

Last fall, we passed OCS legislation that will open resource-rich areas of the Gulf of Mexico. However, there is more we can do. We must pursue a balance of increased efficiency and increased supply.

For example, I would like to re-examine CAFE standards and whether we can be more forward-leaning on fuel efficiency mandates. Among other things, I continue to support authorizing the Administration to increase standards on passenger vehicles.

While energy self-sufficiency is our ultimate goal, energy is—and will remain— a world market. We will always be directly impacted by production and consumption trends in other nations. Thus, it is appropriate that today we will gather information on how to engage our domestic and foreign policy to deal with the world as it is today.

Chapter 3

PREPARED STATEMENT OF HON. GORDON H. SMITH, U.S. SENATOR FROM OREGON

The price and availability of oil and natural gas affects every American. That's why it is important that the Energy Committee's first hearing of the 110th Congress is focused on the global oil situation and its implications for U.S. economic and national security interests. As a nation, we now depend on oil imports to meet sixtypercent of our oil needs. Even modest disruptions in the world supply can result in price spikes at the pump, as we have seen in recent years.

I read the testimony of today's witnesses with great interest, and noticed that— despite widely divergent backgrounds—they all had many recurrent themes. First, there is little surplus production capacity relative to global demand. Much of the current production is controlled by national oil companies that are often making political rather than economic decisions, and are not making the investments needed to maintain and expand production capacity.

Second, much of world's oil is produced in politically unstable nations such as Iraq, Iran, Venezuela and Nigeria. Other nations, such as Russia, are beginning to use their vast energy resources to extract political concessions from their customers. The European Union has recently warned its members against relying too heavilyon one supplier for oil or natural gas.

The global oil market has fundamentally changed as emerging economies such as China and India face growing energy demands. China is now the world's fastest growing oil importer, and is seeking long-term supply contracts around the world. China and the developing nations in Asia are projected to account for 46 percent of the growth in global oil demand by 2030.

Hurricane Katrina showed how vulnerable the United States is to a domestic supply disruption. It also helped us to understand how geographically concentrated U.S. refining capacity has become. All of these factors should lead us to reexamine our energy security strategy. We cannot reduce our dependence on oil without aggressively addressing the transportation sector. Transportation accounts for 70 percent of our nation's oil use, and the transportation sector is almost exclusively fueled byoil.

CAFE standards for automobiles have been stagnant for more than a decade. In 2002, I joined with Senators Kerry and McCain to sponsor an amendment to the energy bill to increase CAFE standards to 36 miles per gallon by 2015. We were told at the time that this would harm the domestic auto industry and reduce consumer choice. Unfortunately, since that time the domestic auto and auto parts industries have lost over 215,000 jobs. Consumers have not benefited either. They lose every time they go to the gas pump.

We were able to use American ingenuity to put a man on the moon almost 40 years ago. We need to use that same spirit of innovation to get a family of five from Portland, Oregon to Yosemite National Park (approximately 740 miles) on one tank of gas. That's why I joined with several of my colleagues in the 109th Congress tointroduce the Fuel Economy Reform Act, which sets a target of an annual increase of four percent a year for fuel efficiency gains. We intend to reintroduce this bill in the near future, and will press for its early consideration.

I believe that the Energy Policy Act of 2005 provided many necessary incentives for the development of renewable energy resources and a new generation of cleaner, more-fuel efficient vehicles. We need to enhance these incentives and send signalsto the investment community about our nation's long-term commitment to renewable resources and to cleaner, more fuel-efficient vehicles.

I look forward to hearing from the witnesses today and to working with my colleagues to address these important energy security issues.

Chapter 4

PREPARED STATEMENT OF HON. BERNARD SANDERS, U.S. SENATOR FROM VERMONT

Good morning. I am pleased to be a member of the Energy and Natural Resources Committee and look forward to the excellent work that this Committee will be doing to ensure a more sane energy policy for our country. Whether it is requiring an increased commitment to renewable sources of energy in the electricity sector or to ensuring appropriate royalty payments from drilling on our public lands, this Committee has a tremendous responsibility.

I sincerely appreciate Chairman Bingaman and Ranking Member Domenici bringing together such an astute panel for the first Energy Committee hearing of the 110th Congress. The geopolitics of oil is a topic that none of us can afford to ignore and while I don't agree with every idea put forward by today's witnesses, I thank them for their time to address us this morning.

What is most striking to me is that, in the prepared testimony, each of the witnesses discusses the dire need to increase efficiency in our transportation sector. I believe—in no uncertain terms—that our failure to increase mileage standards has let the American people down. As consumers look to make each and every dollar go further, they find that, despite the technology being available, their automobiles get the same, or worse—even lower, gas mileage than they did twenty years ago. Additionally, as we grapple with global warming, I believe we must do everything we can to get the most out of each gallon of fuel because the emissions from our cars are simply off the charts. In fact, in Vermont, vehicle emissions are the single largest source of greenhouse gas emissions. I hope, with the help of the witnesses, that we can begin moving forward by starting with a serious discussion of increasing CAFE standards.

Chairman Bingaman, Ranking Member Domenici, I look forward to your leadership on this Committee.

The CHAIRMAN. Thank you very much for your comments, and your statement will be included, obviously, in the record.

Senator Dorgan asked to be recognized for a minute, and if other Senators want to have a minute, they are certainly entitled to do that. This being the first hearing of the year, I do think we need to get to the witnesses fairly soon, but go ahead, Senator Dorgan, with any comments you have.

Chapter 5

STATEMENT OF HON. BYRON L. DORGAN, U.S. SENATOR FROM NORTH DAKOTA

Senator DORGAN. Mr. Chairman, thank you very much. I will be brief, but it is an important hearing. I'm really pleased that you've called it.

There is an old saying, if you don't care where you're going, you'll never be lost. That's important to think about with respect to this country's energy policy.

Oil is critically important. We will always use fossil fuels, always need oil. We suck about 84 million barrels out of the earth a day. Here, in the United States, with our population, we use 1/4 of all the oil that is sucked out of this earth.

We are overly dependent on foreign sources of oil, especially given the national security implications of that dependency, and yet I think we're baby stepping on these issues. We need much more aggressive approaches to reduce that dependency, much more aggressive approaches on renewables.

Let me give you one example. In 1916, we put in place in this country tax incentives, robust permanent tax incentives for the exploration and drilling of oil. 1916, permanent. What have we done for production tax credit for renewables? Well, we put something in place in 1992, short-term. We've let it expire three times. We've extended it five times. We're baby stepping without major commitments and without a decision about where we want to head and how we want to get there.

When we passed the energy bill of 2005, I was proud of it. It moves us down the road, but we need to be much bolder and much, much more aggressive, and I think what we will hear today is about the national security implications of us not doing the right thing and not being bold enough.

So this is a good start, Mr. Chairman. Thank you for allowing me to say a few words and I hope this Congress will give us an opportunity to really be bold and be aggressive on these issues.

The CHAIRMAN. Thank you very much.

Are there other members that wanted to make an opening statement or would you rather go to the witnesses? Anyone else? Senator Salazar.

Chapter 6

STATEMENT OF HON. KEN SALAZAR, U.S. SENATOR FROM COLORADO

Senator SALAZAR. Senator Bingaman, Chairman, and to Senator Domenici as well, I just wanted to say that I very much look forward to the work of this committee in continuing the bipartisan tradition that we worked on for the last 2 years, where we delivered on the 2005 Energy Policy Act and I look forward to working with our new colleagues, Senators Tester, Sanders, Lincoln, Corker and Sessions.

I think that the energy issue, at the end of the day, Mr. Chairman, is one of the very most important issues, perhaps one of the top two issues that face our world today, and I think that this committee has the jurisdiction to help us move forward with the kind of vision and program that Senator Dorgan was talking about. And I look forward to working with my colleagues in achieving that vision.

The CHAIRMAN. All right. Thank you very, very much.

Let me go ahead with our first witness, Dr. Fatih Birol, who is the chief economist and the head of the Economic Analysis Division of the IEA, based in Paris. Thank you very much for being here. Why don't you go right ahead.

Chapter 7

STATEMENT OF DR. FATIH BIROL, CHIEF ECONOMIST, HEAD OF THE ECONOMIC ANALYSIS DIVISION, INTERNATIONAL ENERGY AGENCY, PARIS, FRANCE

Dr. BIROL. Thank you very much, Mr. Chairman. Good morning, ladies and gentlemen. First of all, let me thank you for the kind invitation.

Looking at the next few decades, we think the world is facing twin energy-related threats. One is the increasing risk for energy security and the second one is the energy-related environmental concerns.

For the sake of this meeting, I will focus on the energy security, but I should mention that energy and environmental policies are very much linked to each other. When I'm talking about energy security, we think there are two issues here. Not only oil but natural gas is also an important issue in terms of energy security and plays an important role in the geopolitics of energy as well. But again, for the sake of this meeting, I am going to focus on oil.

We think the oil markets are going through profound changes, which would have a set of implications for our economic, for our domestic policies, as well as for our energy security.

I would like to bring to your attention four aspects of this oil market. The first one is the changes in the demand side; second, in the supply side; the third one, what are the policy implications of these changes; and fourth, what can be done in order to address those policy implications.

On the demand side, first of all, I think the general focus is on the supplier issues, but on the demand side, there is a major change happening.

First of all, the bulk of the demand growth is coming from the developing countries. China and India together are responsible for about 50 percent of the oil demand growth in the last few years, and looking at the future, we expect China and India will be responsible for more than $^{2}/3$ of the oil demand growth.

Why? Very simple. In China today, ladies and gentlemen, 13 persons out of 1,000 persons own a car compared with the United States, where 780 persons out of 1,000 own a car. With the increasing income levels in China, one of the first things they do is to buy a car, which in turn, fuels the oil demand growth.

So the oil demand growth, mainly coming from the developing countries and—and both in developing countries as well as in the OECD countries, will be consumed by the transportation sector— by cars, trucks and jets. And this is a strategic edge to recognize that the bulk of the resources are in OPEC.

Why is it important to note that the bulk of oil will come from the transportation sector? Because in the transportation sector, we do not have readily available alternatives to oil products. You cannot put coal into the tank of your car. And the situation is very different, therefore, from the 1970's and 1980's, where we were using oil for many other purposes, for example, to produce electricity.

But then we hit the price shocks in the 1970's and 1980's. We were able to switch from—in terms of electricity, from oil to nuclear or renewables or gas or other things. But in the case of transportation, there is a concentration, and from the consumers' point of view, our room of maneuver is very, very limited. Therefore, I do recognize that this is the Achilles' heel transportation sector of our current economic system, the lack of concentration on the demand side.

On the supply side, we see concentration as well, the number of producers and who they are. We do recognize, in the energy article of the IEA, that in the next 10 years, none of—the production will come to a peak and afterwards will decline. The difference between the global oil demand worldwide and the decline in—production will need to be made by a very few number of countries where you have the reserves.

Who are those countries? Saudi Arabia, Iran, Iraq—these are three countries which have a lot of potential which could bring the oil to the markets. They have a lot of reserves and it is very cheap to bring that oil to the markets in those countries, and potentially, they are the three countries which can bring oil to the markets.

However, there are very important issues related to this trend. First of all, the number of countries are diminishing. Whoever these countries are, it is always a bit useful to consumers if there is a concentration on the surplus side, that a number of countries are diminishing the suppliers, and they work together in certain decisionmaking processes. Second, most of those countries—all of them—come from a geopolitically unstable region. This is another issue that we should look at very carefully. And the third one is that the enlistment framework in those countries is completely different than we used to see in the past. In the past, we have seen that when the prices went up in the 1970's and 1980's, we received a lot of volumes, oil coming from the North Sea and the Gulf of Mexico, because money went there. The free money went there and as a result of that, production increased and oil came to the markets.

But in this case, in Saudi Arabia, for example, the Saudis will decide how much oil production will grow and the oil purchase may not be the only determinant in their decision how the production capacity will grow.

In the case of Iran, the situation is also similar. Iran desperately needs money in order to invest to increase the production capacity, but Iran does not have—unlike Saudi Arabia—domestic capital. It needs to get money from our side, but the geopolitical context in Iran will make the life very difficult for Iran to increase the production capacity.

So these are the uncertainties, for the consumers: a number of supplies are diminishing that are coming geopolitical, the unstable region, and the investment market is a very new one. There is not free access to capital to those countries.

What are the implications? I will just briefly mention the implications for this concentration on the demand side, on the transportation sector, and concentration on the supplier side, a very few number of countries. First of all, in terms—relative to the process, we should be used to—they should be used to seeing that the process will be volatile and maybe more and more determined by the producers. The producers say, we have a higher rate in the future by looking at this picture.

The risk for the supplier destruction will increase because of the very little number of suppliers, and in addition to that, the leverage of the producers is set to increase with the increase in share.

I wanted to talk a bit on what to do, Mr. Chairman, but I see that I am out of the time. If you allow 2 more minutes, I would like to complete—

The CHAIRMAN. Yes, why don't you go ahead and give us the short version of what we need to know. That would be great.

Dr. BIROL. OK. So I would like to suggest three areas in terms of domestically and internationally. The first one is boosting the domestic production, oil production in the country. Of course, looking at the sensitivities here and there, increasing the efficiency, especially in the transportation sector, which I recognize as the Achilles' heel of the system, and the alternative fuels, in terms of the transportation sector, such as biofuels, will be very important.

Internationally, I imagine that the bulk of the oil is in a very few number of countries, but globally, we are looking at transparence of the reserves data. We are not sure how the reserves data is put together, how much reserve is left in which country and what are the terms there. I think there is a need to put some light, shed some light on the reserves transparency, and finally, perhaps, it may not be a bad idea if one would put some efforts in order to gain some access to the areas where it is legally bound to foreign capital, such as some key Middle East countries, perhaps within the context of WTO. So these are some areas that I wanted to highlight. Thank you.

[The prepared statement of Dr. Birol follows:]

PREPARED STATEMENT OF DR. FATIH BIROL, CHIEF ECONOMIST, INTERNATIONAL ENERGY AGENCY

Mr Chairman, Members of the Committee, it is a privilege to address this Committee on the critical issue of the oil market outlook and its policy implications. The energy future which we are creating is unsustainable. If we continue as before, theenergy supply to meet the needs of the world economy over the coming years will remain too vulnerable to failure arising from sudden supply interruption and will cause serious environmental problems. The oil market is a global one so it is important to provide a global context. To that end, this testimony draws upon the World Energy Outlook 2006, [1] published by the International Energy Agency.

This testimony will examine in turn the outlooks for Demand, Supply and Investment, followed by a look at the potential impact of Alternative Policies and Measures. I would first like to highlight the following key points:

1. *The world is facing twin energy-related threats*: that of not having adequate and secure supplies of energy at affordable prices and that of environmental harm caused by its use. The World Energy Outlook 2006 confirms that fossil-fuel demand and trade flows, and greenhouse-gas emissionswould follow their current unsustainable paths through to 2030 in the absence of new government action—the underlying premise of the Reference Scenario. It also demonstrates, in an Alternative Policy Scenario, that a package of policies and measures that countries around the world are considering would, if implemented, significantly reduce the rate of increase in demand and emissions. Importantly, the economic cost

of these policieswould be more than outweighed by the economic benefits that would come from using and producing energy more efficiently.

2. *Oil demand grows* by 1.3% per year through 2030 in the Reference Scenario, reaching 116 million barrels per day (mb/d) in 2030—up from 84 mb/ d in 2005. The pace of demand growth slackens progressively over the period. More than 70% of the increase in oil demand comes from developingcountries (notably China and India), which see average annual demand growth of 2.5%.
3. *The transport sector absorbs most of the increase in global oil demand.* In the OECD, oil use in other sectors barely increases at all. In developing countries too, transport contributes the bulk of the increase in oil demand. The lack of cost-effective substitutes for oil-based automotive fuels will make oil demand more rigid.
4. *Oil supply is increasingly dominated by a small number of major producers*, most of them in the Middle East, where oil resources are concentrated. Non-OPEC production of conventional crude oil is set to peak within a decade. OPEC's share of global supply grows significantly, from 40% now to 48% by 2030. Iran and Iraq have significant potential to expandtheir production, but Saudi Arabia remains by far the largest producer. The need for more transparent and comprehensive data on oil (and gas) reserves in all regions is a pressing concern.
5. *The oil industry needs to invest* a total of $4.3 trillion (in year-2005 dollars) over the period 2005-2030, or $164 billion per year. The upstream sector accounts for the bulk of this. Almost three-quarters of upstream investments will be required to maintain existing capacity.
6. *A critical uncertainty is whether the substantial investments needed in the oil production sector in key Middle East countries will, in fact, be forthcoming.* These governments could choose deliberately to develop production capacity more slowly than we project in our Reference Scenario. Or external factors such as capital shortages could prevent producers from investing as much in expanding capacity as they would like. As demonstrated by a Deferred Investment Case, slower growth in OPEC oil production drives up the international oil price and, with it, the price of gas.
7. *The new policies analysed in the Alternative Policy Scenario halt the rise in OECD oil imports by 2015.* OECD countries and developing Asia become more dependent on oil imports in 2030 compared to today, but markedly less so than in the Reference Scenario. Global oil demand

reaches 103 mb/d in 2030 in the Alternative Policy Scenario—13 mb/d lower than in the Reference Scenario. Additional policy measures to promote improved fuel ef ficiency of cars and trucks, as well as a greater market share for biofuels, therefore have the effect of improving energy security.

8. *Our analysis demonstrates the urgency with which policy action is required.* Each year of delay in implementing the policies analysed would have a disproportionately larger effect on energy security. Yet there are formidable hurdles to be overcome. It will take considerable political will to push through the policies and measures in the Alternative Policy Scenario, many of which are likely to encounter resistance from some industry and consumer groups. Politicians need to spell out clearly the benefits to the economy and to society as a whole of the proposed measures. In most countries, the public is becoming familiar with the energy-security and environmental advantages of action to encourage more efficient energy use and to boost the share of renewables.

DEMAND

Primary oil demand is expected to continue to grow steadily over the projection period in the Reference Scenario, at an average annual rate of 1.3%. It reaches 99 mb/d in 2015 and 116 mb/d in 2030, up from 84 mb/d in 2005 (Table 1). The pace of demand growth nonetheless slackens progressively, broadly in line with GDP, averaging 1.7% in 2005-2015—only just below the average of the last ten years— and 1.1% in 2015-2030. Preliminary data for 2005 indicate that global oil demand rose by 1.3%—well down on the exceptionally high rate of 4% in 2004.

Most of the increase in oil demand comes from developing countries, where economic growth—the main driver of oil demand—is highest (Figure 1[*]). China and the rest of developing Asia account for 15 mb/d, or 46%, of the 33-mb/d increase in oil use between 2005 and 2030, in line with rapid economic growth. At 3.4% per year on average, China's rate of oil-demand growth is nonetheless below the 5.1% rate of 1980-2004. The Middle East, which experiences the fastest rate of demand growth, accounts for a further 3.8 mb/d. Higher oil revenues than in the last twodecades boost economic activity, incomes and, together with subsidies, demand for oil.

---

[*] All figures have been retained in committee files.

Table 1. World Primary Oil Demand[1] (Million Barrels Per Day]

	1980	2004	2005	2010	2015	2030	2005-2030[2]
OECD	41.9	47.5	47.7	49.8	52.4	55.1	0.6%
North America	21.0	24.8	24.9	26.3	28.2	30.8	0.9%
United States	17.4	20.5	20.6	21.6	23.1	25.0	0.8%
Canada	2.1	2.3	2.3	2.5	2.6	2.8	0.8%
Mexico	1.4	2.0	2.1	2.2	2.4	3.1	1.6%
Europe	14.7	14.5	14.4	14.9	15.4	15.4	0.2%
Pacific	6.2	8.2	8.3	8.6	8.8	8.9	0.3%
Transition economies	8.9	4.3	4.3	4.7	5.0	5.7	1.1%
Russia	n.a.	2.5	2.5	2.7	2.9	3.2	1.0%
Developing countries	11.4	27.2	28.0	33.0	37.9	51.3	2.5%
Developing Asia	4.4	14.2	14.6	17.7	20.6	29.7	2.9%
China	1.9	6.5	6.6	8.4	10.0	15.3	3.4%
India	0.7	2.6	2.6	3.2	3.7	5.4	3.0%
Indonesia	0.4	1.3	1.3	1.4	1.5	2.3	2.4%
Middle East	2.0	5.5	5.8	7.1	8.1	9.7	2.0%
Africa	1.4	2.6	2.7	3.1	3.5	4.9	2.4%
North Africa	0.5	1.3	1.4	1.6	1.8	2.5	2.4%
Latin America	3.5	4.8	4.9	5.1	5.6	7.0	1.5%
Brazil	1.4	2.1	2.1	2.3	2.7	3.5	2.0%
Int. marine bunkers	2.2	3.6	3.6	3.8	3.9	4.3	0.6%
World	64.4	82.5	83.6	91.3	99.3	116.3	1.3%
European Union	n.a.	13.5	13.5	13.9	14.3	14.1	0.2%

[1] Includes stock changes.
[2] Average annual growth rate.
* n.a.: not available.

Demand in OECD countries, especially in Europe and the Pacific region, rises much more slowly. Nonetheless, the absolute increase in North America—5.9 mb/d over the Outlook period—is the second-largest of any region, because it is already by far the largest consumer. The economies of non-OECD countries will remain considerably more oil-intensive, measured by the amount of oil used per unit of grossdomestic product (at market exchange rates), than those of OECD countries. The transport sector absorbs 63% of the increase in global oil demand in 20042030. In the OECD, oil use in other sectors hardly increases at all, declining in power generation and in the residential and services sectors, and growing in industry. Most of the increase in energy demand in non-transport sectors is met by gas, coal, renewables and electricity. In non-OECD countries, too, transport is the biggest contributor to oil-demand growth; but other sectors—notably industry—also see significant growth.

SUPPLY

Resources and Reserves

According to the Oil and Gas Journal, the world's proven reserves [2] of oil (crude oil, natural gas liquids, condensates and non-conventional oil) amounted to 1293 billion barrels [3] at the end of 2005—an increase of 14.8 billion barrels, or 1.2%, over the previous year. Reserves are concentrated in the Middle East and North Africa(MENA), together accounting for 62% of the world total. Saudi Arabia, with the largest reserves of any country, holds a fifth. Of the twenty countries with the largest reserves, seven are in the MENA region (Figure 2). Canada has the least developed reserves, sufficient to sustain current production for more than 200 years. The world's proven reserves, including non-conventional oil, could sustain current production levels for 42 years.

Proven reserves have grown steadily in recent years in volume terms, but have remained broadly flat as a percentage of production. Since 1986, the reserves-to-production, or RIP, ratio has fluctuated within a range of 39 to 43 years. A growing share of the additions to reserves has been coming from revisions to estimates of the reserves in fields already in production or undergoing appraisal, rather than from new discoveries. Some of these revisions have resulted from higher oil-price assumptions, allowing some oil that is known to exist to be reclassified as economically exploitable and, therefore, moved into the proven category. The application of new technology has also improved reservoir management and boosted recovery rates. The amount of oil discovered in new oilfields has fallen sharply over the past four decades, because of reduced exploration activity in regions with the largest reserves and, until recently, a fall in the average size of fields discovered. These factors outweighed an increase in exploration success rates.

Over the past ten years, drilling has been concentrated in North America, a mature producing region with limited potential for new discoveries. Less than 2% of new wildcat wells drilled were in the Middle East, even though the region is thought to hold over 30% of the world's undiscovered crude oil and condensates and is where the average size of new fields discovered in the ten years to 2005 have been higher than anywhere else (Figure 3).

There has been a recent increase in the average size of new discoveries for each new wildcat well drilled, bucking the trend of much of the period 1965-1998. The size of newly discovered fields has continued to decline, largely because exploration and appraisal activity has been focused on existing basins. However, the application of new technology, such as 3D seismic, has increased the discovery success rate perwildcat well, particularly since 1998—boosted by rising global oil demand and a resulting increase in exploration and appraisal activity—and, to a lesser extent, since 1991, with the advent of deep-water exploration (Figure 4). Nonetheless, the average size of discoveries per wildcat well—at around 10 million barrels—remains barely half that of the period 1965-1979. The reduction almost to zero of Middle East exploration, where discoveries had been largest, was the main reason for the lower average size of discoveries since the 1980s.

Exploration and appraisal drilling is expected to increase to offset rising decline rates at existing fields and the consequent need to develop new reservoirs—particularly in MENA, where some of the greatest potential for finding new fields exists. Proven reserves are already larger than the cumulative production needed to meet rising demand until at least 2030. But more oil will need to be added to the provencategory if production is not to peak before then.

Table 2. World Oil Supply (Million barrels per day)

	1980	2000	2005	2010	2015	2030	2005-20301
Non-OPEC	35.2	43.9	48.1	53.4	55.0	57.6	0.7%
Crude oil	32.2	38.1	41.6	45.5	45.4	43.4	0.2%
OECD	14.6	17.2	15.2	13.8	12.4	9.7	¥1.8%
North America	11.8	10.2	9.8	9.4	9.0	7.8	¥0.9%
United States	8.7	5. 8	5.1	5.3	5.0	4.0	¥1.0%
Canada	1.2	1.4	1.4	1.1	0.9	0.8	¥2.2%
Mexico	1.9	3.0	3.3	3.1	3.1	3.0	¥0.5%
Pacific	0.5	0.8	0.5	0.7	0.5	0.4	¥1.2%
Europe	2.4	6.2	4.8	3.8	2.9	1.5	¥4.5%
Transition economies	11.5	7.7	11.4	13.7	14.5	16.4	1.5%
Russia	10.7	6.3	9.2	10.5	10.6	11.1	0.7%
Other	0.8	1.4	2.2	3.3	3.9	5.3	3.6
Developing countries	6.0	13.2	15.1	17.9	18.5	17.4	0.6%
Developing Asia	2.9	5.3	5.9	6.3	6.1	5.0	¥0.6%
China	2.1	3.2	3.6	3.8	3.7	2.8	¥1.0%
India	0.2	0.6	0.7	0.8	0.8	0.6	¥0.2%

Table 2. (Continued)

	1980	2000	2005	2010	2015	2030	2005-20301
Other	0.6	1.4	1.6	1.7	1.6	1.6	0.0%
Latin America	1.5	3.4	3.8	4.8	5.3	5.9	1.8%
Brazil	0.2	1.2	1.6	2.6	3.0	3.5	3.1%
Other	1.3	2.2	2.2	2.2	2.3	2.5	0.5%
Africa	1.2	2.6	3.5	5.2	5.5	4.9	1.4%
North Africa	0.7	0.8	0.6	0.6	0.6	0.7	0.4%
Other Africa	0.5	1.8	2.9	4.6	4.9	4.3	1.6%
Middle East	0.5	2.0	1.9	1.7	1.6	1.4	¥1.1%
NGLs	2.6	4.9	5.1	5.5	5.8	6.8	1.2%
OECD	2.3.	3.7	3.7	4.0	4.1	4.4	0.7%
Transition economies	0.2	0.5	0.5	0.4	0.5	0.6	1.2%
Developing countries	0.1	0.7	0.9	1.1	1.3	1.8	2.7%
Non-conventional oil	0.4	0.9	1.4	2.5	3.7	7.4	7.0%
Canada	0.2	0.6	1.0	2.0	3.0	4.8	6.4%
Others	0.2	0.3	0.4	0.5	0.7	2.7	8.2%
OPEC	28.2	30.9	33.6	35.9	42.0	56.3	2.1%
Crude oil	26.2	27.8	29.1	30.2	34.9	45.7	1.8%
Middle East	17.9	19.5	20.7	22.0	25.7	34.5	2.1%
Saudi Arabia	9.4	8.0	9.1	9.7	11.3	14.6	1.9%
Iran	1.5	3.7	3.9	3.9	4.4	5.2	1.1%
Iraq	2.6	2.6	1.8	2.2	2.8	6.0	4.9%
Kuwait	1.3	1.8	2.1	2.2	1.8	4.0	2.5%
UAE	1.8	2.2	2.5	2.7	3.1	3.8	1.8%
Qatar	0.5	0.7	0.8	0.7	0.7	0.5	-1.9%
Neutral zone[2]	0.8	0.6	0.6	0.5	0.5	0.5	-0.6%
Non-Middle East	8.3	8.3	8.4	8.2	9.1	11.2	1.2%
Algeria	0.9	0.8	1.3	1.1	1.1	0.7	-2.7%
Libya	1.8	1.4	1.6	1.7	1.9	2.7	2.0%
Nigeria	2.1	2.0	2.4	2.5	2.7	3.2	1.2%
Indonesia	1.5	1.2	0.9	0.8	0.8	0.8	-0.8%
Venezuela	2.0	2.9	2.1	2.2	2.8	3.9	2.5%
NGLs	1.8	2.9	4.3	5.4	6.3	9.0	3.0%
Saudi Arabia	0.7	1.0	1.5	1.9	2.0	2.7	2.5%
Iran	0.0	0.1	0.3	0.4	0.6	1.1	4.8%
UAE	0.4	0.4	0.5	0.7	0.9	1.3	3.6%
Algeria	0.1	0.6	0.8	0.9	0.9	0.7	-0.3%
Others	0.6	0.8	1.2	1.5	1.9	3.3	4.1%
Non-conventional	0.0	0.2	0.2	0.3	0.8	1.5	8.8%
Venezuela	0.0	0.1	0.1	0.1	0.2	0.4	5.8%

	1980	2000	2005	2010	2015	2030	2005-20301
Others	0.0	0.1	0.1	0.1	0.6	1.2	10.5%
TOTAL WORLD	64.9	76.5	83.6	0.2	99.3	116.3	1.3%
Crude oil	58.3	66.0	70.8	91.3	80.3	89.1	0.9%
NGLs	4.4	7.8	9.3	75.7	12.2	15.8	2.1%
Non-conventional oil	0.4	1.1	1.6	10.8	4.5	9.0	7.2%
Processing gains	1.7	1.7	1.9	2.8	2.3	2.5	1.2%

[1]Average annual growth rate 2.0.

[2]Neutral Zone production is shared by Saudi Arabia and Kuwait.

According to the US Geological Survey, undiscovered conventional resources that are expected to be economically recoverable could amount to 880 billion barrels (including natural gas liquids, or NGLs) in its mean case (USGS, 2000). Together with reserves growth and proven reserves, remaining ultimately recoverable resources are put at just under 2300 billion barrels. That is more then twice the volume of oil—1080 billion barrels—that has so far been produced. Total non-conventional resources, including oil sands in Canada, extra-heavy oil in Venezuela and shale oil in the United States and several other countries, are thought to amount to at least 1 trillion barrels (WEC, 2004).

Production

In the Reference Scenario, conventional oil production continues to be dominated by a small number of major producers in those countries where oil resources are concentrated. The share of production controlled by members of the Organization of the Petroleum Exporting Countries, particularly in the Middle East, grows significantly [4]. Their collective output of crude oil, NGLs and non-conventional oil grows from 34 mb/d in 2005 to 42 mb/d in 2015 and 56 mb/d in 2030, boosting their share of world oil supply from 40% now to 48% by the end of the Outlook period. Non-OPEC production increases much more slowly, from its current level of 48 mb/d to 55 mb/d in 2015 and 58 mb/d in 2030 (Table 2). Conventional oil accounts for the bulk of the increase in oil supply between 2005 and 2030, but non-conventional resources play an increasingly important role (Figure 5). The projections to 2010 take account of current, sanctioned and planned upstream projects.

Production in OPEC countries, especially in the Middle East, is expected to increase more rapidly than in other regions, because their resources are much larger and their production costs are generally lower. Saudi Arabia remains by far the largest producer of crude oil and NGLs. Its total output of crude and NGLs growsfrom 10.9 mb/d in 2005, to 13.7 mb/d in 2015 and to 17.6 mb/d in 2030 (including Saudi Arabia's half-share of Neutral Zone production). Most of the rest of the increase in OPEC production comes from Iraq, Iran, Kuwait, the United Arab Emirates, Libya and Venezuela. Other OPEC countries struggle to lift output, with production dropping in Qatar, Algeria and Indonesia. These projections are broadly commensurate with proven reserves. OPEC's price and production policies and national policies on developing reserves are extremely uncertain.

Outside OPEC, conventional crude oil production in aggregate is projected to peak by the middle of the next decade and decline thereafter, though this is partly offset by continued growth in output of NGLs (Figure 6). Production in several mature regions, including North America and the North Sea, which has been in steady decline in recent years, stabilises or rebounds in the near term. This reflects several factors,including the restoration of production capacity lost through hurricanes and other technical difficulties, and the impact on increased drilling to boost production in response to recent oil-price increases. But this trend is expected to be short-lived, as relatively high decline rates and rising costs soon drive output back down again.

In the longer term, only Russia, Central Asia, Latin America and sub-Saharan Africa— including Angola and Congo—achieve any significant increases in conventional oil production.

A lack of reliable information on production decline rates makes it difficult to project new gross capacity needs. A high natural decline rate—the speed at which output would decline in the absence of any additional investment to sustain production—increases the need to deploy technology at existing fields to raise recovery rates, to develop new reserves and to make new discoveries.

Our analysis of capacity needs is based on estimates of year-on-year natural decline rates averaged over all currently producing fields in a given country or region. The rates assumed in our analysis vary over time and by location. They range from 2% per year to 11% per year, averaging 8% for the world over the projection period [5]. Rates are generally lowest in regions with the best production prospects and the highest RIP ratios. For OPEC, they range from 2% to 7%. They are highest in mature OECD producing areas, where they average 11%.

The average quality of crude oil produced around the. world is expected to become heavier (lower API gravity) and more sour (higher sulphur content) over the Outlook period [6].

This is driven by several factors, including the continuing decline in production from existing sweet (low-sulphur) crude oilfields, increased output of heavier crude oils in Russia, the Middle East and North Africa (Figure 7), and the projected growth of heavy non-conventional oil output. This trend, together with increasing demand for lighter oil products and increasing fuel-quality standards, is expected to increase the need for investment in upgrading facilities in refineries.

INVESTMENT

Cumulative global investment in the oil sector amounts to about $4.3 trillion (in year-2005 dollars) over the period 2005-2030, or $164 billion per year, in the Reference Scenario. Investment relative to increases in capacity is highest in OECD countries, where unit costs and production decline rates are high compared with most other regions. Projected oil (and gas) investment needs in this Outlook arehigher than in previous editions, largely because of the recent unexpected surge in the cost of materials, equipment and skilled personnel. Unit costs are assumed to fall back somewhat after 2010, as oil-services capacity increases and exploration, development and production technology improves. Upstream investment accounts for 73% of total oil-industry investment.

The required rate of capital spending over the projection period is substantiallyhigher than actual spending in the first half of the current decade, which averaged little more than $100 billion per year. Investment needs increase in each decade of the projection period as existing infrastructure becomes obsolete and demand increases. Our analysis of the spending plans of the world's leading oil and gas companies through to 2010 shows that they expect their spending to be much higher in the second half of the current decade than the first.

Upstream Investment

Upstream oil spending—more than 90% of which is for field development and the rest for exploration—averages $125 billion per year (Figure 8). Three-quarters of this investment is needed to maintain the current level of capacity in the face of natural declines in capacity at producing fields as reserves are depleted. This investment goes to drilling new wells, to working over existing wells at currently producing fields or to developing new fields. In fact, investment needs are far more sensitive to changes in natural decline rates than to the rate of growth of demand for oil.

Downstream Investment

Cumulative investment in oil refining amounts to around $770 billion ($30 billion per year) in the Reference Scenario. These projections include the investment needed to meet demand growth and additional spending on conversion capacity so that existing refineries are able to meet the changing mix of oil-product demand. Tighter fuel-quality standards aimed at mitigating the environmental impact of fuel use are also obliging the refining industry to invest in new quality-enhancement capacity. The required level of refining capacity, allowing for normal maintenance shutdowns, rises from 85 mb/d in 2004 to 117 mb/d in 2030. The largest investments occur inthe Middle East and developing Asia (Figure 9). Most new refineries will be built outside the OECD (see below).

Although investment in oil tankers and inter-regional pipelines makes up a small proportion of total investment needs to 2030, the sum required rises rapidly throughout the projection period, because of the need to replace a large share of the world's aging tanker fleet. Total cumulative capital spending amounts to around$260 billion. Investment in gas-to-liquids plants in 2005-2030 is expected to amount to $100 billion. Most of this investment occurs in the second half of the projection period. Investment in commercial coal-to-liquids plants, mostly in China, is projected to total over $30 billion.

Investment Uncertainties and Challenges

Over the period to 2010, the total amount of investment that will be made in oiland gas infrastructure is known with a reasonable degree of certainty. Investment plans may change in response to sudden changes in market conditions and some projects may be cancelled, delayed or accelerated for various reasons. But the actual gross additions to supply capacity at various points along the oil-supply chain are unlikely to depart much from those projected in this Outlook. However, beyond 2010, there is considerable uncertainty about the prospects for investment, costs andthe rate of capacity additions. The opportunities and incentives for private and publicly-owned companies to invest are particularly uncertain. Environmental policies could increasingly affect opportunities for building upstream and downstream facilities and their cost, especially in OECD countries. In the longer term, technological developments could open up new opportunities for investment and help lower costs.

The availability of capital is unlikely to be a barrier to upstream investment inmost cases. But opportunities and incentives to invest may be. Most privately-owned international oil and gas companies have large cash reserves and are able to borrow at good rates from capital markets when necessary for new projects. But those companies may not be able to invest as much as they would like because of restrictions on their access to oil and gas reserves in many resource-rich countries. Policies on foreign direct investment will be an important factor in determining how much upstream investment occurs and where.

A large proportion of the world's reserves of oil are found in countries where there are restrictions on foreign investment (Figure 10). Three countries—Kuwait, Mexico and Saudi Arabia—remain totally closed to upstream oil investment by foreign companies. Other countries are reasserting state control over the oil industry. Bolivia recently renationalised all its upstream assets. Venezuela effectively renationalised565 kb/d of upstream assets in April 2006, when the state-owned oil company, PdVSA took over 115 kb/d of private production and took a majority stake in 25 marginal fields producing 450 kb/d after the government unilaterally switched service agreements from private to mixed public-private companies. The Russian government has tightened its strategic grip on oil and gas production and exports, effectively ruling out foreign ownership of large fields and keeping some companies, including Transneft, Gazprom and Rosneft, in majority state ownership. Several other countries, including Iran, Algeria and Qatar, limit investment to buy-back or production-sharing deals, whereby control over the reserves remains with the national oil company.

Even where it is in principle possible for international companies to invest, the licensing and fiscal terms or the general business climate may discourage investment. Most resource-rich countries have increased their tax take in the last few years as prices have risen. The stability of the upstream regime is an important factor in oil companies' evaluation of investment opportunities. War or civil conflict may also deter companies from investing. No major oil company has yet decided to invest in Iraq. Geopolitical tensions in other parts of the Middle East and in other regions may discourage or prevent inward investment in upstream developmentsand related LNG and export-pipeline projects.

National oil companies, especially in OPEC countries, have generally increased their capital spending rapidly in recent years in response to dwindling spare capacity and the increased financial incentive from higher international oil prices. But there is no guarantee that future investment in those countries will be large enough to boost capacity sufficiently to meet the projected call on their oil in the longerterm. OPEC producers generally are concerned that overinvestment could lead to a sharp increase in spare capacity and excessive downward pressure on prices. Sharp increases in development costs are adding to the arguments for delaying new upstream projects. For example, two planned GTL plants in Qatar were put on hold by the government in 2005 in response to soaring costs and concerns about the long-term sustainability of production from the North field. An over-cautious approachto investment would result in shortfalls in capacity expansion. Environmental policies and regulations will increasingly affect opportunities for investment in, and the cost of, new oil projects. Many countries have placed restrictions on where drilling can take place because of concerns about the harmful effects on the environment. In the United States, for example, drilling has not been allowed on large swathes of US federal onshore lands—such as the Arctic National WildlifeRefuge (ANWR)—and offshore coastal zones for many years [7]. Even where drilling is allowed, environmental regulations and policies impose restrictions, driving up capital costs and causing delays. The likelihood of further changes in environmental regulations is a major source of uncertainty for investment.

Local public resistance to the siting of large-scale, obtrusive facilities, such as oil refineries and GTL plants, is a major barrier to investment in many countries, especially in the OECD. The not-in-my-backyard (NIMBY) syndrome makes future investments uncertain. It is all but impossible to obtain planning approval for a new refinery in many OECD countries, though capacity expansions at existing sites are still possible. The risk of future liabilities related to site remediation and plant emissions can also discourage investment in oil facilities. The prospect of public opposition may deter oil companies from embarking on controversial projects. Up tonow, NIMBY issues have been less of a barrier in the developing world.

Technological advances offer the prospect of lower finding and production costs for oil and gas, and opening up new opportunities for drilling. But operators often prefer to use proven, older technology on expensive projects to limit the risk of technical problems. This can slow the deployment of new technology, so that it can take decades for innovative technology to be widely deployed, unless the direct cost savings are clearly worth the risk. This was the case with the rotary steerable motor system, which has finally become the norm for drilling oil and gas wells. These systems were initially thought to be less reliable and more expensive, even though they could drill at double or even triple the rate of penetration of previous drilling systems. The slow take-up of technology means that there are still many regions where application of the most advanced technologies available could make a big impact bylowering costs, increasing production and improving recovery factors. For example, horizontal drilling, which increases access to and maximises the recovery of hydrocarbons, is rarely used in Russia.

As well as lowering costs, technology can be used to gain access to reserves in ever more remote and hostile environments—such as arctic regions and deep water—and to increase production and recovery rates. New technology has enabled the subsurface recovery of oil from tar sands using steam-assisted gravity drainage and closely placed twin horizontal wells, while enhanced oil recovery has been made possible by injecting CO_2 into oil wells and by using down-hole electrical pumps, to allow oil to be produced when the reservoir pressure is insufficient to force the oil to the surface.

Although costs have risen sharply in recent years, much of the world's remaining oil can still be produced at costs well below current oil prices. Most major international oil companies continue to use a crude oil price assumption of \$25 to \$35 per barrel in determining the financial viability of new upstream investment. This conservative figure by comparison with current high oil prices partly reflects caution over the technical risks associated with large-scale projects and the uncertainty associated with long lead times and the regulatory environment.

The current wave of upstream oil investment is characterised by a heavy focus on such projects, involving the development of reserves that were discovered in the 1990s or earlier. Unless major new discoveries are made in new locations, the average size of large-scale projects and their share in total upstream investment could fall after the end of the current decade. That could drive up unit costs and, depending on prices and upstream-taxation policies, constrain capital spending. Capital spending may shift towards more technically challenging projects, including those in arctic regions and in ultra-deep water. The uncertainties over unit costs and lead times of such projects add to the uncertainty about upstream investment in the medium to long term.

Implications of Deferred Upstream Investment

In light of the uncertainties described above, we have developed a Deferred Investment Case to analyse how oil markets might evolve if upstream oil investment in OPEC countries over the projection period were to increase much more slowly than in the Reference Scenario. This could result from government decisions to limit budget allocations to national oil companies or other constraints on the industry's ability or willingness to invest in upstream projects. For the purposes of this analysis, it is assumed that upstream oil investment in each OPEC country proportionate to GDP remains broadly constant over the projection period at the estimated level of the first half of the current decade of around 1.3%. This yields a reduction in cumulative OPEC upstream investment in the Deferred

Investment Case vis-a`vis the Reference Scenario of $190 billion, or 25%, over 2005-2030. Upstream investment still grows in absolute terms.

Lower oil investment inevitably results in lower OPEC oil production. This is partially offset by increased non-OPEC production. Higher oil prices encourage this increased investment and production in non-OPEC countries. They also cause oil demand to fall relative to the Reference Scenario. Higher prices for oil and other forms of energy also reduce GDP growth marginally, pushing demand down further. In 2030, the international crude oil price, for which the average IEA import price serves as a proxy, is $19 higher in year-2005 dollars and $33 higher in nominal terms (assuming annual inflation of 2.3%) than in the Reference Scenario—an increase of about 34%.

As a result of higher prices and lower GDP growth, the average annual rate of global oil-demand growth over 2005-2030 falls from 1.3% in the Reference Scenario to 1.1% in the Deferred Investment Case. By 2030, oil demand reaches 109 mb/d— some 7 mb/d, or 6%, less than in the Reference Scenario (Figure 11). This reduction is equal to more than the current oil demand of China. Higher oil prices encourage consumers to switch to other fuels, use fewer energy services and reduce waste. They encourage faster improvements in end-use efficiency. In the transport sector, they also encourage faster deployment of biofuels and other alternative fuels and technologies, such as hybrids. The size of these effects varies among regions. It is highest in non-OECD countries, because the share of non-transport uses in final demand (which is relatively price-elastic) is higher there than in the OECD and because the share of taxes, which blunt the impact on demand of higher international oil prices, is generally lower.

The drop in world oil demand that results from higher prices is accompanied by an equivalent decline in world production in the Deferred Investment Case. Unsurprisingly, OPEC oil production falls sharply in response to much lower investment (Figure 12). Including NGLs, OPEC output is just over 11 mb/d lower in 2030 than in the Reference Scenario, though, at 45 mb/d, it is still nearly 12 mb/d higher than in 2005. OPEC's share of world oil production remains essentially flat at about 40% over the projection period. In the Reference Scenario, the share rises to 48% in 2030.

The fall in OPEC production is largely offset by higher non-OPEC output, which climbs to 64 mb/d—some 4 mb/d higher than in the Reference Scenario and 14 mb/ d higher than in 2005. Higher prices stimulate faster development of conventional and non-conventional reserves in all non-OPEC regions, as marginal fields become more commercial. About 1 mb/d, or 15%, of the increase in non-OPEC output comes from oil-sands in Canada. As a result, the share of non-conventional oil in total world supply increases from 2% in 2005 to more than 9% in 2030, compared with less than 8% in the Reference Scenario.

Alternative Policy Scenario

The Reference Scenario presents a sobering vision of the next two-and-a-half decades, as the major oil-consuming regions—including the United States—become even more reliant on imports, often from distant, unstable parts of the world along routes that are vulnerable to disruption.

In July 2005, G8 leaders, meeting at Gleneagles with the leaders of several major developing countries and heads of international organisations, including the IEA, recognised that current energy trends are unsustainable and pledged themselves to resolute action to combat rising consumption of fossil fuels and related greenhouse-gas emissions. They called upon the IEA to, ''advise on alternative energy scenarios and strategies aimed at a clean, clever, and competitive energy future''. The Alternative Policy Scenario presented in the World Energy Outlook 2006 is a direct response to that request, which the G8 reaffirmed in July 2006 in St. Petersburg.

The Alternative Policy Scenario analyses how far policies and measures currently under discussion [8] can take us in dealing with the grave energy challenges now being faced. Information on more than 1,400 proposed policies and measures has been collected and analysed. Sectoral and regional effects were also analysed in detail, in order to help identify the actions that can work best, quickest and at least cost.

The results of this analysis are clear: First, implementing the policies and measures that governments are currently considering would lead to significantly slower growth in both fossil-fuel demand and CO 2 emissions. Second, new policies and measures would pay for themselves—the financial savings far exceed the initial extra investment cost for consumers.

Demand in the Alternative Policy Scenario

In the Alternative Policy Scenario, the implementation of more aggressive policies and measures significantly curbs the growth in total primary and final energy demand—a reduction of about 10% relative to the Reference Scenario. That saving is roughly equal to the current energy demand of China. Demand still grows, by 37% between 2004 and 2030, but more slowly: 1.2% annually against

1.6% in the Reference Scenario. The reduction in the use of fossil fuels such as oil is even more marked than the reduction in primary energy demand (Figure 13). It results from the introduction of more efficient technologies and switching to carbon-free energy sources. Nonetheless, fossil fuels still account for 77% of primary energy demand by 2030 (compared with 81% in the Reference Scenario).

Global demand for oil in the Alternative Policy Scenario grows on average by 0.9% per year, reaching 103 mb/d in 2030—an increase of 20 mb/d on 2005 levels, but 13 mb/d (11%) lower than in the Reference Scenario. In 2030, the share of oil in total primary energy demand is 32% in the Alternative Policy Scenario, a drop of three percentage points compared to 2004. By 2015, oil demand will be 15% higher than in 2004, compared to 21% in the Reference Scenario. Increased fuel efficiency in new vehicles, together with the faster introduction of alternative fuels and vehicles, accounts for more than half of the oil savings in the Alternative Policy Scenario. Most of the rest comes from savings in oil use in the industry and building sectors.

These savings are equivalent to the current combined production of Saudi Arabia and Iran (Table 3).

Table 3. World Oil Demand In The Alternative Policy Scenario[1] [mb/d]

	2005	2015	2030	2005-2030	Difference versus Reference Scenario in 2030	
					mb/d	%
OECD	47.7	50.7	49.9	0.2	-5.2%	-9.5%
North America	24.9	27.2	27.7	0.4	-3.1%	-10.2. %
United States	20.6	22.4	22.5	0.3	-2.5%	-10.1%
Canada	2.3	2.5	2.5	0.5	-0.2%	-8.2%
Mexico	2.1	2.4	2.7	1.1	-0.4%	-12.7%
Europe	14.4	14.9	13.9	-0.1	-1.4%	-9.3%
Pacific	8.3	8.5	8.2	0.0	-0.7%	-7.6%
Transition economies	4.3	4.7	5.0	0.6	-0.7%	-11.8%
Russia	2.5	2.7	2.9	0.5	-0.4%	-12.2%
Developing countries	28.0	35.6	44.7	1.9	-6.6%	-12.9%
Developing Asia	14.6	19.4	25.8	2.3	-3.9%	-13.2%
China	6.6	9.4	13.1	2.8	-2.2%	-14.5%
India	2.6	3.6	4.8	2.5	-0.6%	-11.3%
Indonesia	1.3	1.5	2.2	2.0	-0.2%	-7.5%
Middle East	5.8	7.7	8.8	1.7	-0.9%	-8.9%
Africa	2.7	3.3	4.2	1.8	-0.7%	-14.4%

	2005	2015	2030	2005-2030	Difference versus Reference Scenario in 2030	
					mb/d	%
Latin America	4.9	5.3	5.9	0.8	-1.1%	-15.8%
Brazil	2.1	2.5	2.9	1.3	-0.6%	-16.0%
Int. marine bunkers	3.6	3.7	3.8	0.2	-0.4%	-9.8%
World	83.6	94.8	103.4	0.9	-129%	-11.1%
European Union	13.5	13.8	12.8	-0.2	-1.3%	-9.5%

[1]Includes stock changes.

[2]Average annual growth rate.

By 2015, demand reaches 95 mb/d, a reduction of almost 5 mb/ d on the Reference Scenario. Measures in the transport sector—notably those that boost the fuel economy of new vehicles—contribute 59% of the savings over the projection period. Increased efficiency in industrial processes accounts for 13%, and fuel switching in the power sector and lower demand from other energy-transformation activities, such as heat plants and refining, for 9%. More efficient residential and commercial oil use makes up the rest.

Supply in the Alternative Policy Scenario

In principle, lower global oil demand in the Alternative Policy Scenario would be expected to result in a lower oil price than in the Reference Scenario. Production in higher-cost fields mainly located in OECD countries, would be reduced, declining even more rapidly after 2010 than in the Reference Scenario. But concerns aboutthe security of supply might encourage OECD and other oil-importing countries to take action to stimulate development of their own oil resources. For example, the UK government is currently considering such policies (DTI, 2006) and the US Congress is considering allowing more offshore oil exploration and giving royalty relief for offshore production. For these reasons, we assumed that oil production in OECD and other net oil-importing countries—as well as the international crude oil price—remain at the same levels as in the Reference Scenario. As a result, the call on oil supply from the net exporting countries is reduced in the Alternative Policy Scenario. OPEC members and major non-OPEC producing regions, including Russia, the Caspian region and west Africa, are most affected (Figure 14). OPEC production reaches 38.8 mb/d in 2015 and 45.1 mb/d in 2030. The average growth of 1.2% per year is just over half the growth in the Reference Scenario. OPEC's share of theglobal oil market rises from the current 40% to nearly 44% in 2030, but this is five percentage points lower than that in the Reference Scenario.

Crude oil production outside OPEC is projected to increase from 50 mb/d in 2005 to 56 mb/d in 2015 and 58.3 mb/d in 2030 (though 1.8 mb/d or 3% lower than in the Reference Scenario). The transition economies are expected to account for half of this increase. Latin America and West Africa account for most of the remainder. Production in OECD countries is expected to decline steadily from 2010 onwards, as in the Reference Scenario. The share of non-conventional oil production in this scenario in 2030, at 8.7%, is an increase of 7.4 mb/d over current levels. The production of biofuels is also expected to increase substantially, especially in oil importing countries. Globally, biofuel production will grow almost 10 times, from 15 Mtoe in 2004 to 147 Mtoe in 2030. Most of the additional growth, over and above Reference Scenario levels, is expected to occur in the United States and the European Union.

The CHAIRMAN. Thank you very, very much.

Linda Stuntz is our next witness. Linda is a partner with Stuntz, Davis and Staffier and has been involved previously with the Department of Energy in a high position and, most recently, was part of the Council on Foreign Relations Task Force that worked up a report on the national security consequences of U.S. oil dependency. Thank you for being here, Linda.

REFERENCES

[1] The World Energy Outlook series is the leading source for medium-to long-term energy market analysis and has achieved widespread international recognition. It is the annual flagship publication of the International Energy Agency. The latest edition was released on 7 November 2006.

[2] Oil that has been discovered and is expected to be economically producible is called a proven reserve. Oil that is thought to exist, and is expected to become economically recoverable, is called a resource. Total resources include existing reserves, ''reserves growth''—increases in the estimated size of reserves as fields are developed and produced—and undiscovered resources. Comparison of reserves and resource assessments is complicated by differences in estimation techniques and assumptions among countries and companies. In particular, assumptions about prices and technology have a major impact on how much oil is deemed to be economically recoverable.

[3] Oil and Gas Journal (19 December 2005). Includes proven oil-sands reserves in Canada.

[4] OPEC is assumed to be willing to meet the portion of global oil demand not met by non-OPEC producers at the prices assumed (see Chapter 1). A special analysis of the effect of lower OPEC investment in upstream capacity is presented at the end of this chapter.

[5] These rates are based on information obtained in consultations with international and national oil companies, oilfield service companies and consultants. Observed decline rates are generally much lower, as they reflect investment to maintain or boost output at existing fields.

[6] However, upstream projects under development may result in a marginal reduction in the sulphur content and a small increase in the API gravity of installed crude oil production capacity in the next five years, according to the IEA's Oil Market Report (12 September 2006).

[7] Projected oil (and gas) investment needs in this Outlook arehigher than in previous editions, largely because of the recent unexpected surge in the cost of materials, equipment and skilled personnel. Unit costs are assumed to fall back somewhat after 2010, as oil-services capacity increases and exploration, development and production technology improves. Upstream investment accounts for 73% of total oil-industry investment.

[8] In mid-2006, Congress was considering a bill to open up 8% of ANWR.

[9] An example for the US would be the implementation of the reform of CAFE standards proposed by the National Highway Traffic Safety Administration.

Chapter 8

STATEMENT OF LINDA G. STUNTZ, ON BEHALF OF A COUNCIL ON FOREIGN RELATIONS INDEPENDENT TASK FORCE

Ms. STUNTZ. Thank you, Mr. Chairman and members of the committee. It is an honor to be before you today to discuss the report prepared by an independent task force organized by the Council on Foreign Relations, released this past October, entitled, as you described, The National Security Consequences of U.S. Oil Dependency. Today, let me highlight four points from this report.

First, you will not find in this report support for the concept of energy independence for this country. As much as I know many of you on both sides of the aisle espouse this, it is, in fact, unrealistic. Barring Draconian measures, the United States will depend on imported oil for a significant fraction of its transportation fuel needs for the next several decades. Moreover, so long as we consume any oil, even if it is produced domestically, we will be affected by what happens in the global oil market, just as corn or other markets of that nature are affected. We cannot wall ourselves off for that market. Our allies are also dependent on this oil.

Therefore, you will find support in what the task force focused on, reducing our dependence on all oil and on better managing the global energy interdependence, which I noted coincidentally, came up in some of my colleagues' testimony on the panel.

One idea of many suggestions in the report is that the International Energy Agency would work perhaps to expand its membership to include new consumers, such as China, who until the early 1990's, were actually oil exporters. That is one of the many changes that has occurred in the global world market.

Second, the constraints on foreign policy caused by energy require greater integration of foreign policy and energy policy. The newspapers this morning and every morning are replete, whether in Asia, Africa, South America or even Europe, with incidents of energy and foreign policy intermingling, yet the task force was unanimous in the view that energy issues have not received sufficient attention in the formulation and implementation of U.S. foreign policy.

Among other things, the task force recommends that an energy directorate be established at the National Security Council, similar to those that exist now, for counter-proliferation defense policy and international economics.

Third, and it was highlighted by Senator Domenici in his opening speech, one of things that I believe has changed since Senator Jackson and I and some of you first began looking at this very difficult challenge of energy security is the increasing role of national oil companies. The reality today is that national oil companies control some 3⁄4 of the world's oil reserves, as best we can tell. ExxonMobile, the largest privately owned oil company in the world, ranks fourteenth, roughly, on the list of proven reserve holders in the world. We are only beginning to come to terms, I would submit, with this development and what it means for world oil markets.

Interestingly, with their access to reserves in other countries more limited, privately-owned oil companies, such as BP, Shell and others, are returning to those areas that remain open to them, including our U.S. Gulf of Mexico and the North Sea. Over time, however, given where petroleum reserves are located and by whom they are controlled, the world will become increasingly dependent on state oil companies to produce the oil that is needed. Some of these are highly efficient. They utilize advanced technology and they conduct their business in a transparent way on commercial terms that we would understand. Many, however, do not, and we have to determine how to deal with them and what their effects will be on the market.

Fourth, in order to address the national security consequences of U.S. oil dependency, we need a comprehensive approach. And this will not be a surprise or news to this committee, but we need it all, we cannot focus on one or the other. We need to increase the efficiency of oil use, primarily in transportation fuels. We need to use alternative fuels. We need to diversify oil supplies, particularly outside the Persian Gulf, which includes in the United States. We need to make oil and gas infrastructure more efficient and secure. And we need to increase the investment in new energy technologies.

The task force considered—and had a lively debate on—increasing the gasoline tax, increasing CAFE requirements and a tradable permit program for gasoline allowances. Again, it will probably be no surprise to you that while the task force unanimously believed we needed to do one or several of these things, we did not have an agreement on which one of these should be pursued.

With respect to alternative fuels, the task force was enthusiastic, in particular, about the possibility of plug-in hybrid electric vehicles using energy generated, in particular, by nuclear power, so that it could deal with some of the emissions issues, which, of course, are the flip side of all these energy issues.

The task force also recommended specifically removing the 54 cents per gallon tariff on imported ethanol so that U.S. consumers may have the benefit of biomass-derived fuels from countries such as Brazil, where ethanol can be produced at a lower cost than in the United States at this time.

In conclusion, I very much hope that the task force work product will be of assistance to the committee as it deals with these important challenges and I look forward to discussing these matters further with you.

[The prepared statement of Ms. Stuntz follows:]

Prepared Statement of Linda G. Stuntz, on Behalf of a Council on Foreign Relations Independent Task Force

Mr. Chairman and Members of the Committee:

It is an honor to appear before you today to discuss the report, ''National Security Consequences of U.S. Oil Dependency,'' released this past October and authored by an independent task force organized by the Council on Foreign Relations. This task force was co-chaired by James Schlesinger and John Deutch, no strangers to this committee. Members of the task force included experts in foreign policy such as Graham Allison, leading economists such as Martin Feldstein, energy experts such as J. Robinson West, business leaders such as Norman Augustine and experienced energy legislators such as former Senate Energy Committee chairman Bennett Johnston. As a veteran of many energy policy battles myself and one who continues to believe (nonetheless) that, working together, we can improve our own energy security and that of our children, it was a privilege for me to participate in this effort.

Every member of the task force has a separate view on what is most important in this report. I do not purport to speak for all of them today—the report does that. I would highlight four points from this report.

First, you will not find support in this report for ''Energy Independence.'' Indeed, the first of several ''Myths'' highlighted by the report is this one. ''Barring draconian measures, the United States will depend on imported oil for a significant fraction of its transportation fuel needs for at least several decades''[1] Moreover, so long as we use ANY oil, we will be subject to world oil market developments and so will our allies. We cannot wall ourselves off from the global oil market, much as we might wish to. Furthermore, policies that attempt to significantly reduce import dependence could dramatically drive up fuel prices. You will find support for reducing our dependence on all oil and on managing better our global energy interdependence, for example, by encouraging the International Energy Agency to work with new major energy consumers such as China and India[2].

Second, the constraints on foreign policy caused by energy require greater integration of foreign policy and energy policy. Whether in Asia, Africa, South America or even Europe, our foreign policy is directly affected by the role of that nation in the global energy marketplace. Yet, the task force was unanimous in the view that energy issues have not received sufficient attention in the formulation and implementation of U.S. foreign policy. Among other things, the task force recommends that an energy directorate be established at the National Security Council, similar to those that exist now for counterproliferation, defense policy and international economics[3]

Third, one of the things that has changed most in global oil markets over the past two decades is the rise of National Oil Companies. The reality today is that National Oil Companies control some three-quarters of the world's oil reserves. Exxon Mobil, the largest privately owned oil company, ranks only 14th on the list of proven reserve owners.[4] We are only beginning to come to terms with this development and what it means for world oil markets, but with their access to reserves elsewhere increasingly limited, privately owned oil companies are returning to those areas that remain open to them, including the U.S. Gulf of Mexico and the North Sea. Over time, given where petroleum reserves are located and by whom they are controlled, the world will become increasingly dependent on state oil companies to produce the oil that is needed.

---

[1] P. 14.
[2] P. 52.
[3] P. 57.
[4] P. 18.

Fourth, in order to address the national security consequences of U.S. oil dependency, we need a comprehensive approach that: 1) increases efficiency of oil use, primarily in transportation fuels; 2) uses alternative fuels; 3) diversifies oil supplies, particularly outside the Persian Gulf, 4) makes the oil and gas infrastructure more efficient and secure; and 5) increases investment in new energy technologies. The task force considered an increased gasoline tax, increase in CAFE requirements and a tradable permit program for gasoline allowances. While the task force unanimously endorsed the adoption of such measures to slow the growth in petroleum consumption, it did not reach agreement on which of these specific measures to employ. With respect to alternative fuels, the task force was enthusiastic about the possibility of ''plug in hybrid'' vehicles, particularly in conjunction with greater use of nuclear power. The task force also recommends removing the $0.54 per gallon tariff on imported ethanol so that U.S. consumers may have the benefit of biomass-derived fuels from countries such as Brazil, where ethanol can be produced at lower cost than in the U.S.[5]

CONCLUSION

It has been my experience that the energy security debate is one particularly afflicted by misinformation and failure to define the problem being addressed. I commend the Committee for seeking the facts regarding the global oil market, our position in that market, what options are available to us in the near and longer term, and what the costs and benefits of those options are. I truly hope that the task force report can be of assistance to you in this effort and would be pleased to try to answer any questions you may have about the report or the matters it addresses.

The CHAIRMAN. Thank you very much for that testimony.

Next is Dr. Robert Hormats, who is vice chairman at Goldman Sachs (International) and has been a witness before this committee, and several Senate committees that I serve, on numerous times. We welcome you back and look forward to your comments.

[5] P. 39.

Chapter 9

STATEMENT OF DR. ROBERT D. HORMATS, VICE CHAIRMAN, GOLDMAN SACHS (INTERNATIONAL)

Dr. HORMATS. Thank you very much, Mr. Chairman and members of the committee. It is a pleasure to be back here.

Let me make just a few points about the situation we face today. First is that we have a history in this country of going through periods of great crisis followed by periods of prolonged complacency and that has caused energy policy to be sort of light switch—on/ off.

I remember my first time coming to this committee. I was economic advisor to Dr. Kissinger in the 1970's when we had the first big oil crisis, with long lines and the Arab oil embargo and high prices, and at that point, we thought the country would rally behind a very bold policy. Well, there were some major elements of progress in the 1975 Act under President Ford, where we had energy efficiency standards for cars. And after that, a lot of companies moved away from oil to burn other things to generate power and we used coal in other things, as opposed to oil, in the industrial part of the economy.

But then we had other crises that followed, periods of concern, periods where people would say, we need bold policy. Then prices go down and we relax and forget about it. I think that point Senator Dorgan made a moment ago is that even now, we're just taking baby steps. These steps are not commensurate with the situation we find ourselves in today. And that is, people say, look, after 9/11, everything changed. Well, energy policy really has not changed very much. We're fighting a war on terrorism. We are spending money, lots of money, for oil. We're heavily dependent on countries that are very unreliable suppliers. A large portion of money is spent by us and other importers, and goes to countries whose interests are hostile to those of the United States. Some of that money finds its way into terrorist hands. We should accept the fact that that is the case.

So what we're doing now is we're fighting in a post-9/11 environment with a pre-9/11 energy policy. It is simply not sufficient to deal with the national security crisis that we face today. The crisis is a geopolitical one and the vulnerability of this country to disruptions—look what is happening in Nigeria today, kidnappings of people on these oil rigs. We have Venezuela making very tough statements about further nationalization. We have Russia using oil as a political lever. We have instability in the Middle East. If Iran deteriorates further in the relationship—that will affect oil. It has happened before. If Iraq deteriorates further and the civil war increases and other countries start getting involved, then you have additional tensions. If you have tensions between the Shiites on one side of the Persian Gulf and the Sunni on the other, that's going to make transportation of oil all the more vulnerable. And therefore, we have to come up with a much bolder set of energy policies for national security reasons.

The thing about it is, it may be difficult to deal with some of these situations abroad, of which we have less control. We have it within the capability of the United States not to become energy independent. I think Linda has made a very good point: energy independence, at this point, is not possible, but we can manage our vulnerability a lot better than we are doing today and it's the vulnerability that is the huge problem.

How do we do that? We have the capability, for instance, by insisting on tougher fuel standards for automobiles, to improve the efficiency with which we use oil. And it's quite possible to do. It's within the realm of technological possibility. Now there may be reasons why you can't go as fast as we would like, but there should be the target of much greater energy fuel—oil fuel efficiency standards with off-ramps in case there is an economic reason it can't be done or a technological reason it can't be done. There ought to be some way—exceptions for a period of time, but the goal ought to be to reduce the efficiency—to improve the efficiency of the use of oil as a transportation fuel because, by and large, in this country, oil is a transportation fuel and if we can't address that issue, we're not going to address the overall vulnerability issue.

The second part of this is that there are a number of other elements there and the laws that are written and the regulations that come out of this Congress, many of them are too short-term. Many of the incentives have very narrow windows so companies can't take advantage of them. I've listed in my testimony some of these. These are, I think, very important.

Second, some of these structures for the investments that we want to come into this sector are limited so that certain kinds of investors—individual investors in certain cases, institutional investors in other cases—cannot really put money in because of the structure of the law and the way the regulations are written. These are additional important points.

Third, we've got to work with other countries, like China and other countries that are big consumers and growing consumers, not to have a big fight with China over energy, but we ought to try to find ways where we can use our ability to develop, for instance, clean coal technology to help other countries to utilize their energy resources more efficiently and in a way which is environmentally sound.

The last point, generally—and I'll stop because I'm over my time—is we have the technological capability on the supply side. I've talked about the demand side, greater efficiency in the use of oil for automobiles and transportation vehicles. We have the capability, with the entrepreneurialism and the vitality of this country, as President Ford said in 1975 when he first talked about this, to utilize this capability in this country, to develop new, alternative sources of energy. And there was a lot going on. There is a lot more that can be done with the right kind of government incentives.

And we also have to use the conventional sources of energy that we have. We have a lot of them. Environmental practices have improved a lot. We can't do it only by reducing the use of energy. We can't do it only by increasing the supply. But when you combine the two policies that increase our dependence on our own conventional and non-conventional resources, combined with greater efficiency of the use of oil in particular, we have the ability to reduce our vulnerability quite substantially.

But we need a much bolder policy. It can't be a pre-9/11 policy, it has to be a post-9/11 policy. We're in a war and oil is one of the elements of that. During World War II, the American people were called on to buy bonds. There were bond rallies. People asked, what can I do? How can I help? Now, what can Americans do? What they can do is support bold energy policy. If we are committed to a really tough policy, people ought to be supporting efforts to reduce the wasteful use of energy across the board. Thank you.

[The prepared statement of Dr. Hormats follows:]

PREPARED STATEMENT OF ROBERT D. HORMATS, VICE CHAIRMAN, GOLDMAN SACHS (INTERNATIONAL)

Mr. Chairman and Members of the Committee:

Thank you for your kind invitation to testify on the critically important subject of the economic and national security implications of America's oil dependence.

I speak to you today as a citizen concerned about our nation's increasing dependence on potentially unstable supplies of foreign oil. This dependence creates profound economic, political and security vulnerabilities. Also, a portion of the large amounts of petrodollars accumulated by a number of suppliers is used in ways that threaten American interests.

By way of background, I was economic advisor to Dr. Henry Kissinger on the National Security Council staff in the mid 1970s when this country experienced its first energy crisis after the 1973 Yom Kippur War, and participated in his Middle Eastern shuttle diplomacy during the period that followed. At that time, I had high hopes that the Arab oil embargo, the sharp increase in the price of oil, and the long-lines at gas stations would produce a bipartisan consensus on energy policy and jolt our nation into a bold and effective effort to reduce oil dependence and future vulnerability. Indeed, some progress was made. The Energy Policy Conservation Act of 1975, championed by our late President Gerald R. Ford, launched a number of bold initiatives to achieve this goal. And the country did accomplish significant improvements in the efficiency of oil use through compulsory mileage standards for automobiles and because U.S. industry and power plants shifted dramatically away from using oil as a fuel.

But when prices fell later in the decade, a sense of complacency set in. Then we were hit by another crisis that caused oil prices to spike at the end of the 1970s; that was triggered by the fall of the Shah of Iran and the Iranian Revolution. Complacency set in once again after that crisis receded and prices fell. Another oil crisis occurred in 1990 when Iraq invaded Kuwait, after which the sense of urgency about dramatic alterations in energy policy and use faded again. Decade after decade our dependence on foreign oil has risen. In the mid-1970s, 35% of this nation's oil consumption was supplied by imports. Now, three decades later, it is 60%.

After 9/11, again at the beginning of the current Iraq War in 2003, and again during the large price run-up in and the summer of 2006 the country had excellent opportunities and powerful incentives to confront energy vulnerabilities with a bold policy response. The 2005 Energy Policy Act contained a number of positive features—but these measures were not commensurate with the seriousness or the urgency of the energy challenge this country faces.

American dependence on potentially vulnerable oil supplies continues to grow, with little prospect that it will change—despite the fact that we are engaged in a War on Terrorism in which oil imports by the U.S. and other nations provide funds to nations hostile to the U.S. and countries friendly to us. *It is often said that ''9/ 11 changed everything!'' Sadly, in the area of energy policy it hasn't changed very much. American oil vulnerability continues unabated.*

There are several national economic and security consequences of this situation:

- If the situation in Iraq continues to deteriorate and other oil producing nations become more involved, the risks increase to oil supplies not only from disruptions in Iraq but also from greater tensions between the Sunni nations on the western side of the Persian Gulf and the Shiites on the eastern side, with oil facilities and shipments becoming increasingly vulnerable. Moreover, addedwestern pressures on Iran over its nuclear program could lead to oil disruptions or threats thereof;
- The American economy remains highly vulnerable to supply disruptions in oil exporting nations; these could result from acts or terrorism, political instability, efforts to use oil as leverage, or natural calamities;
- High oil prices resulting from strong demand from countries such as the U.S.and other major importers give countries such as Iran and Venezuela added resources to take actions inimical to American interests;
- Oil-dependent friends and allies feel more vulnerable to the pressures and potential use of oil leverage from supplying countries and therefore are reluctant to side with the U.S. on key issues affecting those suppliers;
- Oil-related tensions and competition are likely to intensify—as countries suchas China seek to lock up scarce supplies or make political deals to solidify long-term supply relationships, or suppliers such as Russian and Iran use oil as leverage to extract political concessions from consumers.

My concerns about this untenable and dangerous situation led me—together with a group of other concerned citizens to join the Energy Security Leadership Council in an effort to press for greater and more resolute national action on this matter—and for an end to the divisive, highly polarized debate that has stymied genuine progress on many fundamental issues. The Council, a project of Securing America's Future Energy (SAFE), is a nonpartisan group of business executives and retired military leaders. It recently unveiled a report entitled

''Recommendations to the Nation on Reducing U.S. Oil Dependence.'' (I will discuss a few of these later in my testimony, along with a number of recommendations that I believe can also contribute to progress in this area.) The members of the Council believe that America's energy security is in a perilous state. Along with my fellow Council members, I am convinced that America's leaders must move quickly and steadfastly to confront our high level of oil dependence as a profound national security challenge.

ENERGY INTERDEPENDENCE

Calls for ''energy independence'' offer a false promise to the American people. They also suggest a sort of xenophobia that implies that the U.S. can or should attempt to solve its energy problems with little regard for those of other nations. In fact, oil is a fungible global commodity, which means that events affecting supply or demand anywhere will affect oil consumers everywhere. A country's exposure toworld oil prices or oil price shocks is a function of the amount and types of oil it consumes; the ratio of ''domestic'' to ''imported'' oil is only a portion of the problem. Even if the U.S. could substitute domestic energy for all foreign oil—a goal the Council believes to be impossible—American economic prosperity would still be linked to the health of a global economy dependent on international oil flows. So as we work to enhance our own energy security we should also be strengtheninginternational cooperation with oil producers and consumers to improve global energy security, efficiency, and environmentally responsible production and usage.

It is also important to make a distinction between dependence and vulnerability. There are numerous suppliers of oil that are very reliable and that use the funds earned in a constructive fashion. There are others whose facilities are vulnerable to disruption and that use funds in ways inimical to U.S. interests. But a large portion of the world's oil comes from this in the second category, posing a series of economic, political and security risks.

KEY RECOMMENDATIONS

The Council recommends a goal of cutting U.S. oil intensity the amount of oil it takes to produce a given amount of GDP—in half by 2030. There is a favorable precedent for this objective. Since the mid-1970s, the U.S. has managed to trim oil intensity by 50%, chiefly by raising the fuel efficiency of passenger cars, virtually eliminating oil as a fuel for electric power generation, and expanding less energy-demanding sectors of the economy, particularly in the area of services. As a consequence, the U.S. now uses half the amount of oil to produce a dollar of GDP, in real terms, as it did just thirty years ago. Unfortunately, however, progress in this area has slowed in the last ten years.

One key goal must be to make America's prosperity less dependent on a commodity the production level of which responds only very slowly to changes in price. Combine this price inelastic supply with 1) the vulnerability of oil supplies to various types of disruption, 2) the fact that some countries see oil as a political as well as an economic commodity, and 3) the fact that much of the world's production is in the hands of state owned oil companies, many of which use oil revenue for political or social ends rather than reinvest it in new production capacity, and you have the recipe for severe energy-related economic disorder.

As a result of the halving of U.S. oil intensity since the mid-1970s, the high oil prices experienced in the summer of 2006 represented a smaller relative cost to the economy than in the past. Further reductions in oil intensity would provide a measure of insurance against some of the effects of sudden future oil price shocks or sustained high oil prices. In addition, by boosting the production of alternative sources of energy to displace oil, we can create more production capacity at home, keep more of our money in this country, create a great number of new, high quality jobs in industries that manufacture and utilize new environmentally responsible energy production and conservation technologies, and develop new export products that can be sold to an energy hungry and environmentally conscious world.

The Global Energy Challenge

The global economy is in the midst of a period of extraordinary growth that promises to transform the lives of billions of people, bringing comforts and luxuries to regions where humankind has long struggled for subsistence. Creativity and the drive for a better life are the engines of this tremendous surge in output, living standards and productivity but like all engines they require energy to function.

By 2020, world energy demand is forecast to jump by 50% over 2000 levels, with most of the increase coming in developing countries. The safe and affordable delivery of all this energy is by no means assured. Even if resources turn out to be sufficient in the aggregate, their distribution will not map closely to the topography of demand. The resultant uncertainty of supply and upward pricing pressure will exacerbate international tensions stemming from non-energy issues.

Oil provides only 40% of global energy, but, as the premier transportation fuel, it has emerged as the touchstone of the world's energy outlook. On both economic and psychological grounds, oil price spikes threaten the prosperity of many nations, including many of the poorest on this planet. They also sow the seeds of tension between exporting and importing nations, among consuming nations, and among different groups within countries. Indeed, since so much oil is used for personal transportation, oil prices have an enormous impact on the pocketbooks of virtually every American family. Correspondingly, policy efforts that impact oil's cost and availability must take into account the interests of the average American family and quickly become major political issues.

America's Clear and Present Dangers

For much of the last century, surplus domestic oil production reduced U.S. vulnerability to oil disruptions elsewhere in the world. But America's oil production is now dwarfed by current consumption. Thus, while the U.S. remains the third largest oil producer in the world, domestic production can satisfy barely 40% of its requirements.

The U.S. generates 28% of the world's goods and services while consuming roughly a quarter of its oil production. This may seem like a balanced, even favorable energy equation, but closer inspection reveals a different story. Despite considerable progress toward more efficient energy use, America requires substantially more oil to create a dollar of Gross Domestic Product (GDP) than is the case in most other developed countries. Some of this differential in "oil intensity" can be attributed to our nation's vast size, the dispersion of our population, and less reliance on public transportation. Global military obligations, which are inextricably linked to our commitment to secure the flow of oil for the benefit of all nations, further increase American consumption. But even with these extenuating factors, there can be little doubt that the U.S. can and must use energy far more efficiently.

America's long-term supply and demand balance is no more encouraging. U.S. oil demand is expected to grow 24% over the next two decades, and even if new discoveries raise its current 3% share of global oil reserves, our nation will almost certainly still require substantial amounts of petroleum imports. Import dependence will also define energy security for our key allies and most of the world's manufacturing nations. Unfortunately, the developed nations that consume most of the world's oil are not in a good position to produce the fuels they need.

A large portion of the world's oil reserves are owned by state-owned or controlled oil companies in non-O.E.C.D. countries. It is worth underscoring this point—especially because when oil prices were rising last summer there were many accusations, misguided in my view, that this was a conspiracy among the big oil majors, when in fact the *six largest state oil companies have ten times the reserves of the top six privately owned companies.* Some of these state companies are highly efficient and well run, but others are highly politicized and are not able to utilize their profits to increase production or modernize capacity. Because of the large state company role in the world's oil markets, there is not a ''free market'' for oil. As a result, a substantial portion of production is politically influenced and production decisions and practices are frequently economically suboptimal.

With each passing year, the global oil trends now at work—rising consumption, reduced spare production capacity, politicized spending decisions, and potentially high levels of instability in key exporting countries—all increase the likelihood of an energy crisis. The odds in favor of a crisis are further heightened by the rise ofterrorist movements bent on targeting critical elements of the world's vulnerable oil production, processing, and delivery infrastructure.

Given today's precarious balance between oil supply and demand, taking even a small amount of oil off the market could cause prices to rise dramatically. In Oil Shockwave, a cabinet-level oil crisis simulation conducted in 2005 by SAFE and the National Commission on Energy Policy (NCEP), a 4% global shortfall in daily oilsupply only 3.5 million barrels in a 84 million barrel daily market resulted in a 177% increase in the price of oil, to over $150 per barrel. The simulation was played out by men and women who have served in the highest ranks of the U.S. government; Robert M. Gates, our current Secretary of Defense, for example, filled the role of National Security Advisor. The hypothetical scenarios put before the participants were designed to simulate a decline in world oil production due to regional instability and to terrorism. The incidents were completely plausible, and some, such as unrest in Nigeria, have subsequently come to pass. But there was little these skilled officials could do to stop a gut-wrenching increase in the price of oil. Indeed, one of the major lessons of the simulation was that the Strategic Petroleum Reserve (SPR), the emergency supply of federally owned crude oil, offers only very limited protection against a major supply disruption. Emergency reserves cannot sustainthe United States through a prolonged crisis, and it will be extremely difficult to reach political consensus on when it is appropriate to begin using them.

Even under normal conditions, oil dependence has severe economic consequences. In 2005, direct outlays for imported oil accounted for a third of the country's $800 billion current account deficit. In 2006 prices, these outlays have gone still higher. By diverting funds away from domestic consumption and investment, oil imports put a drag on U.S. economic growth and undercut the nation's long-term competitive position. Oil dependence also adds billions to our defense expenditures by making overseas protection of oil supplies a high strategic priority.

China

Before I turn to a discussion of recommendations, I want to touch upon the rise of China and how that impacts U.S. energy security. Some observers have insisted that clashes between the U.S. and China over energy are inevitable. Chinese companies are buying oil properties and concluding long-term supply contracts around the world. A few of China's deals are in countries such as Venezuela, Iran and Sudan, with which the U.S. has strained or no relations. Also, China's surge in oil demand was seen, incorrectly, by some as a reason for higher prices last summer. And China's increased coal production concerns U.S. environmentalists.

But the fact is that the U.S. and China have many common interests in the energy area and thus many reasons to cooperate. Consider these facts:

- The U.S. is the world's biggest oil importer. China is the world's fastest growing oil importer. High prices and supply instability harm both nations. Price increases in the summer of 2006 primarily reflected the lingering affects of sluggish world investment in production and refining in the previous decade, and market perception of high political risk that could disrupt oil deliveries, which both nations have an interest in correcting.
- Chinese, like Americans, are concerned about their environment. China faces colossal and urgent environmental problems, as anyone who has visited Beijing during the winter has experienced first hand. U.S. companies have great expertise in clean energy technology that could help.
- The U.S. and China have a similar interest in open sea lanes for oil.

- Both nations also desire a secure business and legal environment for their energy investments in emerging economies as well as stable and growing supplies from world exporters.

When I look at China and the U.S., I see two nations that have an enormous interest in cooperation in pursuit of energy security. Several areas are ripe for a common effort.

- A Joint Business-Government Commission on Clean Coal Technology; this could help China develop and utilize its massive amounts of coal in an environmentally responsible way and boost U.S. exports of technology and equipment in this area.
- Joint research on alternative fuels, which should also include experts from Japan, would draw on the best talent in these three countries. This could lead to breakthroughs in, or broader dissemination of, non-carbon based production and use technologies.
- Strengthen U.S.-China cooperation in the context of the International Energy Agency.
- Consultation with one another, and with other regional nations, to maintain open sea lanes, that could reassure China that the U.S. will not use its naval power to leverage China on oil.
- Strengthen established regional groups that include China, the U.S., and other Pacific nations to address environmental and energy supply issues.

Helping China to increase domestic energy output using state-of-the-art, environmentally responsible, technologies would slow the growth of its oil import dependence, reduce imbalances in global markets, dampen global price pressures, and contain the process of global warming. And cooperation on these broad energy issues can strengthen broader U.S.-China ties. The alternative—frequent energy confrontations—benefits neither country.

There Are No Silver-Bullet Solutions

Success in improving the nation's energy security posture will demand significant public and private investment—supported by meaningful tax and other non-tax incentives like loan guarantees—over a sustained period. Because of the volatility of markets and the strategic role of oil, a considerable amount of government support is needed to provide the necessary incentives through a supportive and reliable regulatory, and tax environment if we are able i) to reduce America's oil vulnerability; ii) strengthen this nation's capacity to produce oil and alternative sources of energy; and iii) utilize energy more efficiently and in an environmentally responsible way. The U.S. is capable of major breakthroughs if all elements of our society work together.

The good news is that we Americans have it within our power and our technological and financial capacity to take meaningful steps to reduce oil dependence and increase energy security using both proven methods and technologies and our ingenuity and entrepreneurial capacity to develop new breakthroughs.

- *Improving efficiency:* In the view of the Council, the most important thing the U.S. can do to lessen its oil dependence in the near and medium-term is to utilize oil considerably more efficiently. With the goal of once again halving oil intensity—as in the 1980s and 1990s—in the space of two decades, Americans can do much to protect the economy against the effects of oil shocks that can be unleashed by forces beyond our control. Improved vehicle fuel efficiency is the single most important avenue for further cutting the nation's oil intensity.
- *We must face the hard fact that in the U S. oil is primarily a transportation fuel; unless we can dramatically curb the use of oil in our cars and trucks, we will be unable to reduce our oil dependence.* Currently the direction is not positive; through 2030 oil usage by SUVs and light duty trucks is expected to surge by roughly 77%. The transportation sector accounts for nearly 70% of all the oil the country uses; and oil fuels almost 97% of all transportation. With most of the vehicles on the nation's roads operating at efficiency levels far below what is achievable with currently available technologies, there is a clear opportunity to realize sizable fuel economy gains without overall loss of safety or functional utility. We propose empowering the National Highway Traffic Safety Administration (NHTSA) to mandate annual fuel efficiency increases of 4% while allowing for these increases to be postponed or constrained if economic, technical, or safety impediments are demonstrated.
- *Increasing stable supply.* As a second means of improving America's oil risk profile, the Council recommends efforts to increase the production of oil in stable regions of the world, including in the U.S., Canada, and Mexico. We must move beyond the drill/don't drill debate for this simple reason: by facilitating the discovery and recovery of conventional oil

resources, in conjunction with stricter environmental standards, American investment and the capabilities of this nation's formidable oil experts and oil service companies can ease the tight supply conditions that unsettle oil prices and lessen the probability that even modest supply shortfalls will trigger an international oil crisis. By the same token, a robust nuclear power program also makes great sense.

- *Supplies abroad.* Just as significantly, by working to ensure the rule of law, sanctity of contracts, and stable investment climates abroad, America can help to lower the likelihood of future disruptions. There a great many potential projects that can enable the world to diversify the sourcing of oil away from its present growing concentration on the Middle East. By utilizing groups such asthe International Energy Forum—a ministerial dialogue among major energy producers and consumers established in 2003—the conditions for increased investment in such projects can be enhanced.
- *Developing alternatives.* Third, America can lead the way in expanding the availability of alternatives to petroleum-based fuels. Diversifying our transportation fuel supply must be a key part of any comprehensive effort to improve

U.S. energy security. Without an expanded supply of alternatives, conventional petroleum will continue to power nearly all of our motor transport. Such reliance on a single non-substitutable input creates profound economic dangers. To date, through the help of federal policies such as the Renewable Fuels Standard, the phase-out of MBTE as an additive, and tax incentives, corn-based ethanol has developed as one of the most successful domestic alternative transportation fuels. Production in the United States rose from 1.4 billion gallons a year in 1995 to about 4 billion in 2005.

Although this growth rate is impressive, it is merely a drop in the bucket when compared to this nation's annual gasoline consumption of 140 billion gallons; it is equivalent in terms of energy content to only 2% of our gasoline demand. At a maximum, corn-based ethanol may be able to displace 10% of our gasoline use before corn demand outstrips supply. Consequently if we want to have a significant impact on reducing our consumption of petroleum-based fuels, the federal government must encourage the development and commercialization not only of dedicated energy producing crops such as corn and sugar, but also of other potentially large-volume bio-fuels like cellulosic ethanol (which are low cost, do not compete with the food chain, and provide another revenue stream for farmers) that is generated from forest residue and agricultural waste such as wheat straw, switch grass, and corn stover. Technologies like cellulosic ethanol are poised to dramatically raise bio-fuels production by shifting acreage-to output ratios.

However, to transform this promise into reality, existing federal policies, like the federal loan guarantee program for innovative technologies must be fully funded and implemented; and new policies, which encourage and support investment in commercial facilities and related transportation infrastructure must be readily adopted.

There are two specific policy changes that I believe would enhance the development and commercialization of renewable energy.

The first is for *Congress and the Administration to take a longer term per spective in the way tax incentives are structured.* For example, with respect to the production tax credit (PTC) for renewable energy sources like wind and geothermal, the credit is available to projects that are placed in service before January 1, 2009. Historically this credit has been renewed only for short periods of time and often after great uncertainty and delay. This on-again, off-again process has added significant uncertainty to such projects and increased their costs. Therefore a longer-term extension of the PTC, say for five years would help to avoid such problems. The second is to *alter the structure of tax incentives to encourage investment from additional categories of investors, including small investors, by enabling them to benefit from tax incentives—thereby increasing the avail ability and lowering the cost of capital for these projects.* For example, the way the law is now written retail investors and taxpayers paying the alternative minimum tax cannot use the production tax credit for investment in wind projects; allowing the use of master limited partnerships for these types of projects would broaden the group of investors who could help to finance them.

It is worth noting that many of the new technologies being developed involve high technical risk, significant costs, and regulatory uncertainties—and that costs of demonstration projects to show that these technologies can be deployed on a commercial scale as well as those associated with their commercial development are significantly greater than the initial R&D costs. Therefore, maximizing the range of investors supplying capital, providing reliable incentives, and creating and funding policies that reduce the significant financial risks associated with these projects are critical to advancing the process of proving and commercializing innovative energy alternatives.

- *Managing risks:* In the Council's view, we must manage risk within the interdependent global oil economy. In our dangerous world, threats are one commodity not in short supply. America contributes far more than any other nation to protecting this global infrastructure, and the time has come for other nations to expand their own efforts. All nations have an interest in the stability of the global oil infrastructure, and a variety of international efforts could help to ensure the smooth flow of oil. New multilateral accords should play a role, but there are also opportunities for expanded reliance on existing organizations such as the Gulf Coordination Council, NATO, or ASEAN. A common interest in "oil security maintenance" in partnership with producing nations offers real potential to improve regional security in areas of rising geopolitical competition by creating frameworks for pragmatic international cooperation. Where appropriate, the U.S. should provide exporting countries with diplomatic support as well as with counter-terrorism training and other military aid.

I urge you to review the Council's detailed recommendations, which are contained in our published report. We will be glad to provide any further clarifications you may require.

The Capabilities of the American People

Last week, the nation mourned the passing of our 3 8th President, Gerald R. Ford—for whom I had the great privilege of working. President Ford left a legacy of honesty, integrity and decisiveness. These aspects of his leadership were particularly evident in his handling of energy security. In his 1975 State of the Union address, President Ford recognized the energy dangers threatening the country. He expressed a "very deep belief in America's capabilities,"—its innovative capacity and technological skills to overcome its growing dependence on imported oil. He also rallied support for fuel efficiency standards. I share President Ford's optimism in the capacity of Americans to respond to the challenge of growing energy dependence, and his belief that Americans will rally around tougher energy measures, if they are given strong leadership.

America has a long history of pulling together in the face of national security challenges. I am currently completing a book entitled *The Price of Liberty: How America Pays for its Wars.* In all the major national security challenges of the twentieth century, Americans demonstrated a remarkable willingness to make patriotic wartime sacrifices. During World War I and World War II, American's not only paid dramatically higher taxes but also participated in massive bond drives to mobilize billions of dollars to support out troops. Roosevelt's Secretary of the Treasury Henry Morgenthau, when asked about the significance of such drives, said that they were launched not only to raise massive amounts of funds, but also to respond to people who asked ''What can I do to help.'' Today, the answer to this question lies not in buying more bonds but in buying less gasoline.

Since 9/11 there have been no major bond drives as in past wars—and only limited steps to reduce our dependence on oil. The time has come to recognize that energy security is central to the national security challenges of twenty-first century, and to present the American people with the unvarnished truth regarding how oil affects the struggle in which we are engaged. We must meet the threats we face in the same spirit as our parents and grandparents during past wars—with farsighted patriotism and willingness to compromise narrow partisan, ideological, philosophical and economic positions in the long-tern national interest.

There are enormous dangers in facing the challenges of a post-9/11 world with a pre-9/11 approach to energy that relies so heavily on oil from some of the most vulnerable areas of the world and sustains price levels that benefit countries such as Iran and Venezuela that seek to undermine our interests and threaten our friends. American leaders and the American people have rallied the country in past wars; the challenge is to do so again,

I thank you again for this opportunity to testify.

The CHAIRMAN. Thank you very much.

Next is General Charles Wald, who is the former deputy commander of the U.S. European Command, and he has been very involved with the Energy Security Leadership Council. We very much appreciate his being here today to give us his views.

Chapter 10

STATEMENT OF GENERAL CHARLES WALD, U.S. AIR FORCE [RETIRED], FORMER DEPUTY COMMANDER, U.S. EUROPEAN COMMAND, AND MEMBER, ENERGY SECURITY LEADERSHIP COUNCIL

General WALD. Thank you, Mr. Chairman and members of the committee, for the opportunity to discuss the global oil balance and its impact on U.S. and national security.

I recently retired from the Air Force after 35 years of service and during my career had the opportunity to fly combat over Vietnam, Cambodia, Iraq and Bosnia and learned much regarding how to use military assets to effectively solve national security problems.

But I also learned that many believed the U.S. military is solely responsible for security. I like to call this the ''Dial 1-800-The-U.S.Military'' syndrome, because it reflects how people assume the U.S. military is a toll-free resource that can be called on to perform tasks that no one else has either the capability or will to execute.

I recall a recent meeting with several major global oil company executives in Kazakhstan. Before we began our discussion, one of the executives thanked me and the U.S. military for protecting the free flow of oil around the world. The executive's world view included the expectation that the U.S. military will be there to provide worldwide security and to ensure the free flow of oil without any assistance from others. This struck me, and frankly, does not seem like a good model, particularly for the United States. The U.S. cannot and should not be everywhere to protect all the vulnerable components of the global oil infrastructure.

With regard to the oil dependence issue, military response and capabilities are by no means the only effective tools available and in many cases are not appropriate. In fact, the single most effective step the United States can take to improve its energy security is to increase transportation efficiency.

The transportation sector is responsible for nearly 70 percent of the oil the United States consumes. CAFE standards legislated in 1973 during the Arab oil embargo were instrumental in helping America lower oil usage by the 1980's, but there has been little progress since the original mileage targets were met.

As a consequence, America's light-duty vehicle fleet now has the worst average fuel efficiency in the developed world. America must do better in this area. That is why, as you mentioned, the Energy Security Leadership Council has recommended vehicle efficiency standards that require 4 percent annual improvements in mileage per gallon performance but with regulatory off-ramps.

The National Transportation Safety Administration finds the 4percent requirement to be technically infeasible, unsafe or not cost effective.

Some may be surprised to hear from a former General talk about fuel efficiency standards but they shouldn't be. In the military, we learned that forced protection isn't only about protecting weak spots, it's also about reducing vulnerabilities before you go into harm's way. That's why lowering the Nation's demand for oil is so critical.

Nearly all of our U.S. military commands have some oil security tasks and in essence they provide a blanket of security that benefits all nations. Central Command guards access to the oil supplies in the Middle East; Southern Command defends Colombia's Cano Limon pipeline; Pacific Command patrols the tanker routes in the Indian Ocean, the South China Sea and the Western Pacific; and my last assignment, as deputy commander of European Command, which included, by the way, most of Africa. We patrolled the Mediterranean, provided security in the Caspian Sea and off the West Coast of Africa.

During that assignment, I became more appreciative of the size and scope of the oil security challenge. While surveying that challenge, it became apparent that the U.S. military could not protect that vast infrastructure without partners—and trust me, there should be partners in this mission. The free flow is clearly in the best interests of people all over the world. These interested parties certainly cannot replicate all the capabilities of the U.S. military, but their contributions can free up military tasks that only the U.S. military can successfully accomplish.

That's why we created a program to train local populations and militaries in the Caspian region, to develop effective policing capabilities. That's also why we worked and engaged international oil firms in creating programs for protecting our assets. At the end of the day, this cooperative government, industry and military approach is the only realistic way to address the growing vulnerability of our worldwide supply.

The armed forces of the United States have thus far been successful in fulfilling our energy security mission and they continue to carry out their duties professionally and with great courage. As a result of this success, many have come to believe—and I believe, falsely—that energy security can be achieved solely by military means. We need to change this paradigm because the U.S. military is not the best instrument for confronting all the strategic dangers emanating from oil dependence. The 1973 oil embargo is the most famous example of the use of energy as a political strategic weapon.

Currently, it has been Russia that has shown the most willingness to use oil as a political tool. At the beginning of 2006, Russia suspended natural gas exports to the Ukraine, which, in turn, withheld natural gas destined for Western Europe. Again, just this week, Russia has stopped natural gas exports to Belarus, with much the same effect as the 2006 event.

In an oil-dependent world facing increasingly tight supplies, the growing power of oil exporting countries and the shift in strategic calculations of other important countries have all added up to lessen U.S. diplomatic leverage.

Iran, which exports to the United States' European and Asian allies, has threatened the use of the oil weapon to retaliate against efforts to constrain their nuclear program. The European Union relies on Middle Eastern oil, and Russian gas continues to complicate U.S. foreign policy efforts, especially when considering our efforts to stop Iran from developing nuclear weapons. China, with its rapidly growing dependence on foreign oil also blocks U.S. diplomatic initiatives in an effort to strengthen its own ties with oil exporters.

Given all these factors, it is imperative that the United States make energy security a top strategic priority. Toward that end, we should mobilize and leverage all of our national security resources, including our economic power, our investment markets, our technological products and our unsurpassed military strength.

I've mentioned many specific recommendations in my written testimony. One recommendation I would like to mention here is to call for the U.S. Government to reorganize its bureaucracy to better address the needs of a comprehensive international energy strategy and I recommend the Department of Defense, as was mentioned earlier for the National Security Council, to designate an individual as their energy security policy expert and director.

In sum, we need a comprehensive national security strategy for energy security. We must be prepared for sudden supply shocks triggered by terrorism or politics. We must promote greater diversity of fuel options while improving the efficiency of our Nation's fleet. Most of all, we must have the courage to shape the future rather than to succumb to the paralysis of resignation. It is time for America to lead the way in constraining oil consumption and boosting stable oil production, to work with other nations to secure its production of energy products and to maintain the military resources that will continue to be essential for ensuring energy security.

These are not easy tasks. Making progress will take enlightened and courageous leadership. I thank you all for the opportunity to discuss this important issue to our national security.

[The prepared statement of General Wald follows:]

Prepared Statement of General Charles F. Wald, USAF (Ret.), Former Deputy Commander, U.S. European Command, and Member, Energy Security Leadership Council

Chairman Bingaman and Members of the Committee, I thank you for inviting me to talk to you about the global oil balance and its impact on U.S. economic and national security. My friend Bob Hormats has done a great job describing the work of the Energy Security Leadership Council on which we both serve, and let me say that I agree completely with his assessment that oil dependence is one of the most serious economic and national security challenges facing this nation.

Please allow me the opportunity to explain. I retired from the U.S. Air Force last July after thirty-five years of service. During my career I flew combat missions over Vietnam, Cambodia, Iraq, and Bosnia, and I learned a lot about how to use military assets to effectively solve national security problems. But I also learned that a lot of people think the military is solely responsible for national security. I like to call this the ''Dial 1-800-The U.S. Military'' syndrome, because it reflects how people assume the military is a ''toll-free'' resource that can be called on to perform tasks that no one else has the capability for or the will to execute. I remember once I was introduced to some oil company executives in Kazakhstan, and before we began talking one of them thanked me and the U.S. military for protecting the flow of oil around the world. He was serious and sincere about this, and I was seriously concerned. This man's world view included the expectation that the U.S. military will be there to provide security all over the world to ensure the free flow of oil without assistance from others. It did not seem like a good model to me.

And it's not just a matter of cost, though this approach does burden the military's budgets as well as its personnel and their families. It is really an issue of recognizing that true national and economic security must rest on a nation's full strength, not just on the backs of its military. This is necessary because some threats cannot be mastered through military means. In the case of the oil dependence problem, military responses are by no means the only effective security measures, and in some case are no help at all.

In fact, the single most effective step the U.S. can take to improve its energy security is to increase transportation efficiency. The transportation sector is responsible for nearly 70% of all the oil the country consumes. Within the transportation sector, oil—nearly 13 million barrels per day of it—accounts for 97% of delivered energy. More than 8 mb/d are used to fuel the over 220 million light-duty vehicles that Americans rely on for mobility.

For thirty years, these vehicles have been subject to government-mandated Corporate Average Fuel Economy (CAFE) standards enacted in the aftermath of the 1973-1974 Arab oil embargo. CAFE was instrumental in helping Americans lower oil usage by the early 1980s, but unfortunately its requirements for cars have remained essentially unaltered since they were put in place and those for light trucks have been improved only slightly. As a consequence, America's light-duty vehicle fleet now has the lowest average fuel efficiency in the developed world.

America must do better in this area, and that is why the Energy Security Leadership Council has recommended vehicle efficiency standards that require 4% annual improvements in miles per gallon performance, but with regulatory "off-ramps" to protect manufactures and consumers if analysis by the National Highway Traffic Safety Administration (NHSTA) finds 4% to be technically infeasible, unsafe, or not cost-effective for a given year.

Some in the audience may be surprised to hear a former general talk about fuel efficiency standards, but they shouldn't be. In the military you learn that force protection isn't just about protecting weak spots, it's about reducing vulnerabilities before you get into harm's way. That's why lowering the nation's demand for oil is so important. If we can lessen the oil intensity of our economy, making each dollar of GDP less dependent on petroleum, that will mean we're less vulnerable if andwhen our enemies do manage to successfully attack elements of the global oil infrastructure.

The best ways to reduce oil intensity are to improve efficiency and to develop the ability to produce and use realistic amounts of alternative fuels such as ethanol.

Political forces have often portrayed increased supply and decreased demand as mutually exclusive ambitions.

As I have been saying, the U.S. needs a comprehensive policy for achieving genuine energy security. This policy should include increases in production in places like the Outer Continental Shelf along with strict new environmental protections. Increased production in the U.S. makes economic sense, since it will reduce the risk premium that currently inflates the global price of oil.

Last but not least, they are energy security tasks that must involve the military, acting either alone or with partners around the globe. I'd now like to offer a bit of background in that area.

The Military's Historical Involvement in Energy Security

The United States protects the global oil trade for the benefit of all nations. In part, this is because the U.S. has unmatched military capabilities. But another reason is that other nations know the U.S. military is out there doing the job.

The implicit strategic and tactical demands of protecting the global trade have been recognized by national security officials for decades, but it took the Carter Doctrine of 1980, proclaimed in response to the Soviet Union's invasion of Afghanistan, to formalize this critical military commitment.

The Carter Doctrine committed the U.S. to defending the Persian Gulf against aggression by any ''outside force.'' President Reagan built on this foundation by creating a military command in the Gulf and ordering the U.S. Navy to protect Kuwaiti oil tankers during the Iran-Iraq War.

The Gulf War of 1991, which saw the United States lead a coalition of nations in ousting Iraqi leader Saddam Hussein from Kuwait, was an expression of an implicit corollary of the Carter Doctrine: the U.S. would not allow Persian Gulf oil to be dominated by a radical regime—even an 'inside force' that posed a dangerous threat to the international order.

More recently, the security agenda in the Gulf has expanded beyond state actor aggression to include concerns about terrorist attacks on facilities and supply lines.

Threats Abound

Since issuing his 1996 ''Declaration of War'' against the U.S. and its partners, Osama bin Ladin has warned of attacks on oil installations in the Persian Gulf. Last year, the world came close to experiencing an oil supply shock when an Al-Qaeda attack on the Abqaiq facility through which approximately 60% of Saudi Arabian oil exports pass was barely foiled. In addition to attacking physical infrastructure, Al Qaeda operatives have also targeted expatriates in their residential areas, in particular in Riyadh, Saudi Arabia (October 2002) and in al-Khobar (May 2004).

Iraq is also the scene of persistent insurgent and terrorist attacks on pipelines and pumping stations, especially in the North of the country. These attacks have severely limited Iraqi oil exports to the Mediterranean through Turkey, and they are a major reason why Iraqi oil production has stubbornly remained below its prewar peak. The lost output has cost Iraq billions of dollars at a time when it needs every dollar and while U.S. taxpayers have spent billions on the reconstruction of the country. But if violence continues, and especially if it spreads to the south,where most of the oil and export facilities are located, then all of Iraq's oil production could be at risk. The implications of this supply cut would be severe.

The danger of attacks on shipping is proven—in October 2002, the French supertanker Limburg was rammed by a small boat packed with explosives off the coast of Yemen. Most oil shipments have to pass through a handful of maritime chokepoints. Roughly 80% of Middle East oil exports pass through the Strait ofHormuz (17 mb/d), Bab el Mandeb (3 mb/d), or the Suez Canal/Sumed Pipeline (3.8 mb/d). Another 11.7 mb/d pass through the Straight of Malacca and 3.1 mb/d through the Turkish Straits. All of these passageways are vulnerable to accidents, piracy, and terrorism. Since alternative routes are lacking, the effect of a major blockage at one of these points could be devastating. Even unsuccessful attacks on tankers are likely to raise insurance rates and thus oil prices.

Partnering for Preparedness

Nearly all of our U.S. military commands handle oil security tasks. Central Command guards access to oil supplies in the Middle East. Southern Command defends Colombia's Cano Limon pipeline. Pacific Command patrols tanker routes in the Indian Ocean, the South China Sea, and the Western Pacific. European

Command is involved in oil security all the way from the Caspian Sea to West Africa.

I happen to know more about European Command, because, in late 2002, I was named its Deputy Chief. It was during this period of my service that I came to realize that I came face to face with the size and scope of the oil security challenge. The global economy relies on a massive oil infrastructure that stretches far beyond the Persian Gulf to pipelines in the Caucasus and offshore drilling rigs in the Gulf of Guinea. Surveying this situation, I realized that the U.S. military could not protect this vast infrastructure without partners. And, trust me, there should be partners out there, because the free flow of oil is in the best interest of many people all over the world. These interested parties certainly cannot replicate all the capabilities of the U.S. military, but their contributions can free up the military for tasks that only it can complete. That's why I made an effort to train local populations in the Caspian region to develop effective policing capabilities. It's also why I work to engage oil industry firms in protecting their assets. At the end of the day, this improved division of responsibilities will benefit the U.S. our Allies, and millions around the world.

Military Power Has Limits

The armed forces of the United States have been extraordinarily successful in fulfilling their energy security missions, and they continue to carry out their duties with great professionalism and courage. But, ironically, this very success may have weakened the nation's strategic posture by allowing America's political leaders and the American public to believe that energy security can be achieved by military means alone. We need to change the paradigm, because the U.S. military is not the best instrument for confronting all of the strategic dangers emanating from oil dependence. This is particularly true when oil is used a political weapon.

The 1973 Arab embargo is still the most famous example of the use of energy as a political strategic weapon. But in recent years, it has been Russia that has shown the most willingness to play this dangerous game, as at the beginning of 2006, when it stopped natural gas exports to the Ukraine, which in turn withheld the natural gas destined for Western Europe. The danger of conflict with a nuclear power like Russia should make it abundantly clear that there are limits on how we can use military power to guarantee energy flows. But we can take political steps to counter Russia's brandishing oil and natural gas as political weapons. Russia wants to join the World Trade Organization (WTO) as a full member. Russia's entry into this organization must be made contingent on its behavior. Russia must make a commitment to fostering energy security; there should be no reward for sowing insecurity.

Of course, energy exporting governments don't need to resort to full-fledged embargoes to hurt the U.S. and other importers. Exporters can manipulate price through less drastic production cuts. Tellingly, after oil prices dropped from their 2006 peak of $78 to about $60 in the U.S. market, OPEC members began to cut back on production. Governments in oil-producing countries can also constrain future supply through investment decisions that lead to long-term stagnant or glowing growth in production and exports, or even decline. Often enough, future supply destruction is the unintended or accepted consequence of an insistence on government control of natural resources. Currently, an estimated 80-90% of global oil reserves are controlled by national oil companies (NOCs), which are highly susceptible to being constrained by political objectives, even if these undermine long-term supply growth. With this level of state-control, it's impossible to speak of a free market for oil.

State-controlled production is frequently inefficient, relying on outdated technology and reserve management techniques. Consider the case of Venezuela. The demagoguery of Hugo Chavez led to a strike at that country's national oil company (PDVSA) in the winter of 2002-2003 and then to the dismissal of thousands of well-trained petroleum engineers. Deprived of the services of expert personnel, Venezuela may suffer the permanent loss of hundreds of thousands of barrels per day of production. Chavez also worsened the financial terms for international oil companies operating in Venezuela, making it even less likely that emerging best practices will be employed in the country's oil fields.

While major international oil companies and their advanced technology still maintain major stakes in Venezuela's oil industry, this is not the case in Russia, whose government has made it abundantly clear that it wants to maintain near absolute control over its energy resources. This power grab has curtailed foreign investment, and ultimately limited production as well.

Russia's oil industry stands as a testament to the dangers of political meddling in oil production. After the collapse of the Soviet Union, Russian production plummeted to only 6 mb/d in the mid-1990s, but then the efforts of private companies helped push production back to over 9 mb/d, achieving 10% annual growth rates in 2003 and 2004 [1]. However, with the subsequent expropriations of private enterprises such as Yukos, the production growth curve has flattened. Government control over production in Russia will also adversely impact the massive Shtokman natural gas field and Sakhlain-2 oil projects. President Putin has determined that tight government control of resources is more important than the greater revenue that would accrue from increased production achieved through cooperation with Western oil companies.

Oil Constrains U.S. Foreign Policy

In an oil-dependent world facing increasingly tight supplies, the growing power of the oil-exporting countries and the shifting strategic calculations of other importing countries have lessened U.S. diplomatic leverage.

Iran, which exports to the U.S.'s European and Asian allies, has threatened to use the ''oil weapon'' to retaliate against efforts to constrain the country's nuclear program. Russia's growing self-assurance and assertiveness cannot be divorced from the leverage it enjoys because of its oil and gas resources.

European Union reliance on Middle Eastern oil and Russian gas continues to complicate U.S. foreign policy efforts, especially as far as efforts to stop Iran from developing nuclear weapons are concerned. China, with its rapidly growing dependence on foreign oil, also blocks U.S. diplomatic initiatives in order to strengthen its own ties with oil exporters. Chinese opposition has helped thwart U.N. Security Council sanctions against Iran and prevented significant intervention in the Darfur region of Sudan.

Confronting Diverse Dangers

Giving all these factors, it is imperative that the U.S. make energy security a top strategic priority. Toward that end, we should mobilize all of our national security resources, including our economic power, our investment markets, our technology prowess, and our unsurpassed military strength. To borrow a metaphor from the energy sector, this broad approach will result in some dry-holes, but it should pay solid dividends over time.

The U.S. can set an example with domestic actions. Curtailing demand is the most important security step we can take. But we should also demonstrate a willingness to increase domestic production in an environmentally-responsible fashion. The U.S. should also impress upon other major exporting countries that they need to more fully develop their oil and gas reserves. To enhance the global market's ability to respond to price signals and increase the reliability of global production, access to U.S. markets and global trade organizations should be contingent upon the granting of reciprocal access to foreign investment in energy production. Such access should then be protected by appropriate laws, regulations, and judicial systems that preserve the sanctity of contracts. In keeping with this reciprocity requirement, the U.S. must not take a protectionist stance when foreign nations seek to invest in U.S. oil companies, unless clear national security risks can be demonstrated.

The U.S. should also encourage greater transparency and private ownership in the world's NOCs. This might mean promoting the creation of national oil companies in countries where oil ministries run energy operations, and promoting private majority ownership in those countries where state-run companies already exist. In the long-run, such efforts will depoliticize the decision-making process of oil investing and should lead to more exploration of oil in response to market demand.

Iraq is the oil-exporting nation with which we have the most influence. It is also the country that could boost its oil production and exports most significantly in the medium and long-term (given some political stability). In fact, Iraq, the least developed OPEC country, has the potential to expand its oil production from the current 2.2 mb/d to 6-8 mb/d over the next decade. It is also a country that desperately requires, and is eager for, foreign investment in its energy sector. The U.S. should encourage Iraq to take the steps necessary for increasing production. These steps should include improved energy infrastructure security efforts, increased capital expenditure in the energy sector, a viable Petroleum Law that will encourage necessary foreign investment, and a reconstituted NOC that more effectively excludes political considerations from its operations so as to boost operational efficiency.

The U.S. government should also work with other governments to minimize the likelihood and impact of supply disruptions. In this respect, the U.S. should promote greater security of the global oil infrastructure, which includes everything from ports and tankers to well and pipelines. The keys to infrastructure security are protection, repair, and redundancy. This mission will require an expansion of contingency planning. Multilateral military and civilian rapid response teams should be formed to respond to attacks and repair damage. This will likely involve a good deal of American training of other countries' military and civilian agencies. It will also require the stockpiling of expensive spare parts in key strategic locations around the world. The U.S. and its allies should consider adding energy infrastructure protection as a role for NATO, for instance. Oil companies also need to be fully engaged in such an endeavor with funding and dedicated personnel.

Arranging for oil-exporting nations to store more oil in or near major consuming nations whether in tankers, tanks or petroleum reserves—can serve as a way to minimize the impact of a supply disruption. The oil-producing countries could retain absolute control over that oil, including deciding when to release it and to keep profits from it. The is not a new idea; the U.S. Government and Saudi Arabia have at times raised the idea of storing Saudi oil in the United States, though the details were never worked out [2].

Among consuming nations, the U.S. should promote the build-up of strategic reserves in key locations across the globe. China and India are making some progress in this regard, but only very slowly; indeed, they are planning on building reserve capacity for only 15-20 days worth of imports, while the United States Government's Strategic Petroleum Reserve (SPR) now contains 700 million barrels, or the equivalent of about 60 days of imports. But building reserves is only half the task. There must be clear decision-making processes for when to use these reserves. These processes must be developed both domestically and internationally. Without clear release procedures, strategic reserves cannot offer maximum protection.

One final recommendation that merits mention is a call for the U.S. government to reorganize its bureaucracy to suit the needs of a comprehensive international energy strategy. For example, in the Department of Defense, which has a salient role to play in global oil security, there is no civilian office that is dedicated to coordinating the efforts and needs of the military commands with respect to energy matters. An Office of Energy Security should be formed to do that.

In sum, a comprehensive national security strategy must address numerous energy security issues. We must be prepared for sudden supply shocks triggered by terrorism or political action. We must also be ready to deal with the stagnation of global production and the increasing politicization of the global oil market. We must promote greater diversity of fuel options while improving the efficiency of our nation's vehicle fleet. Most of all, we must have the courage to shape the future rather to succumb to the paralysis of resignation. It is time for America

1. to lead the way in constraining oil consumption and boosting oil production;
2. to work with other nations to secure the production and flow of all energy products; and
3. to maintain the military resources that will continue to be essential for ensuring energy security.

I thank you.

The CHAIRMAN. Thank you very much.

Our final witness this morning is Dr. Flynt Leverett, who is a senior fellow and director of the Geopolitics of Energy Initiative at the New America Foundation and also a visiting professor of Political Science at MIT.

Thank you for being here.

REFERENCES

[1] EIA, ''Country Analysis Brief: Russia,'' (January 2006), available online at www.eia.doe.gov/cabs/Russia/Full.html.

[2] Patrick Clawson and Simon Henderson, ''Reducing Vulnerability to Middle East Energy Shocks: A Key Element in Strengthening U.S. Energy Security,'' Washington Institute for Near East Policy, policy focus #49 (November 2005).

Chapter 11

STATEMENT OF DR. FLYNT LEVERETT, SENIOR FELLOW AND DIRECTOR, GEOPOLITICS OF ENERGY INITIATIVE, NEW AMERICA FOUNDATION, WASHINGTON, DC

Dr. LEVERETT. Thank you, Senator Bingaman, Senator Domenici and members of the committee, for the chance to speak with you this morning. I will try to get to the heart of the topic for this hearing, namely the geopolitics of oil.

I will start with a very stark assessment and that is, in my view, during the next quarter century, the most profound challenges to America's continued global leadership will flow from the strategic and political consequences of the structural shifts in global energy markets that previous witnesses have been laying out for you.

And in the time that I have, I'd like to talk with you about what I see as those strategic and political consequences of the structural shifts in the global oil market and what they mean for American interests.

On both the supply and demand side of the global oil market, we have seen strategic and political responses to the kinds of structural shifts that Dr. Birol and others have described for you.

On the supply side, we've seen the rise of what a lot of folks call ''resource nationalism''. Resource nationalism is often defined as national government with oil and gas resources asserting their ownership rights over those resources in ways that work against the interests of international energy companies, something like

Mr. Chavez's recent declaration about nationalizing projects to develop extra heavy crude in the Orinoco region.

But there is another dimension to resource nationalism that I think is very important here and that is the use by energy suppliers of their status as suppliers in a tight market as a source of political leverage. Venezuela is a good example in this hemisphere, obviously Russia is an important example, but there are many others that you could lay out that are very important for American interests. Saudi Arabia, for example, using its unique status as the swing producer in the world oil market to cultivate a kind of alternative strategic partnership with China, as a hedge against a further deterioration in its traditional strategic partnership with the United States. This phenomenon, this aspect of resource nationalism will, I think, pose an increasingly serious set of challenges to American interests in coming years.

On the demand side, we see an analogous phenomenon, what I describe as "resource mercantilism", namely the reliance of energy importing states, China and India being the outstanding examples, on national energy companies to secure access to overseas oil and gas resources on a more privileged basis than simple supply contracts.

In terms of the significance of this phenomenon, I'm not particularly concerned about Chinese and Indian state-owned energy companies locking up some critical mass of oil and gas reserves and keeping them off of international markets. Today, the equity oil produced by the Chinese national energy company abroad amounts to less than .5 percent of all the oil that is produced in the world today. Even with the most optimistic assumptions of how many equity oil deals Chinese energy companies will be able to conclude around the world by 2020, you're still talking about no more than 2 percent of the oil that is going to be produced in the world.

It's not so much the market impact, but it's really the geopolitical impact that is important. As Chinese and Indian state-owned energy companies go about pursuing these deals with the support of their governments, it basically puts these countries into competition for geopolitical influence with the United States and puts us into competition for influence with these countries in very strategically important regions—the Middle East, central Asia, Sub-Saharan Africa—and if present trends continue unchecked, this is going to become an increasingly important source of geopolitical tension around the world and an increasingly important source of challenges for U.S. interests.

Now, resource nationalism and resource mercantilism pose significant challenges to American interests, each in its own way, but I would also point out that these two phenomena can intersect in some particularly challenging ways for the United States.

One of the ways in which they intersect is in what I have described as a ''new axis of oil'', namely a loose coalition of states— energy-producing states and energy-importing states, loosely organized around a Sino-Russian axis. This axis of oil is bolstering Sino-Russian cooperation on a whole host of strategic issues and I believe this axis of oil is emerging as the principle counterweight to American hegemony in global affairs.

Let me give you a couple of examples of what I mean. The axis of oil, this Sino-Russian axis of oil, has been quite successful over the last 2 to 3 years in essentially rolling back the projection of U.S. influence into central Asia following the September 11 terrorist attacks. Russia and China have cooperated in standing up the Shanghai Cooperation Organization, the world's largest regional security organization and the only such organization in the world in which the United States is not a participant. Working together in the Shanghai Cooperation Organization, Russia and China have basically been able to lock us out of central Asia.

Another example of the way the axis of oil is working against American influence is the Iranian nuclear issue. As other witnesses have suggested, it is Chinese and Russian collaboration, particularly in the Security Council, but in other arenas as well, that is frustrating a very significant segment of U.S. policy objectives on the Iranian nuclear issue.

Let me stick with Iran for a minute because I think that the geopolitical and geo-economic stakes go well beyond the nuclear issue, as important as that is. There is, I would argue, a broader strategic competition underway between the United States on the one hand and Russia and China on the other concerning Iran's economic and political role in the Middle East and global energy markets in coming decades. Essentially, the outcome of this competition hinges on which countries will assume leading roles in helping Iran develop its hydrocarbon resources.

Iran's resource base is truly impressive. If you take its gas reserves—the second largest in the world—convert them into barrels of oil equivalent, and add them to their oil reserves—also the world's second largest—you basically have a situation in which the aggregate hydrocarbon reserves of Iran and the aggregate hydrocarbon reserves of Saudi Arabia are effectively the same. And each of those countries is significantly larger in terms of aggregate hydrocarbon reserves than Russia.

What this means, given Iran's low rate of production, is that Iran is basically the only major energy producing country in the world that has the resource potential to increase its production of both oil and natural gas by orders of magnitude in coming decades. But to do that, Iran is going to have to get a lot of investment and a lot of technology transfer.

U.S. policy bars U.S. energy companies from participating in that process. We threaten secondary sanctions against European companies that would consider participating in those projects. And all of that, combined with pretty lousy investment policies inside Iran, has had an effect. There has been serious under-investment in uranium production capacity. Some have even suggested that given the level of under-investment, Iran is going to face a serious crunch in terms of its ability to meet its domestic needs and continue exporting in a significant way in coming years.

But that assumes that Iran doesn't have an alternative and Iranian officials and energy executives will tell you that, increasingly, they think they do have an alternative. It's a three-pronged alternative. It means taking advantage when there are European companies that are prepared to do deals in Iran. Fine, let them do the deals. Second, it involves developing strategic partnerships with— let me call them generically third-world energy companies. Chinese energy companies are at the top of that list and Chinese energy companies are increasingly committing to put very, very significant sums of investment capital into the development of Iranian oil and gas. Then the third prong of this strategy, the most recent prong, is cooperation with Russia, particularly to help Iran develop its potential as a gas exporter. In return for that help, Iran has agreed to—the phrase that is used is to coordinate the marketing of its gas exports to ensure that Iran's emergence as a major gas exporter does not work against Russian interests. This is the strategy that the Iranians think will get them out of the box that they are in now. The United States and even Europe, to a large extent, are basically irrelevant to that strategy.

The potential for Russian and Chinese cooperation to develop Iran's hydrocarbon resources, I think, the potential for that cooperation and its impact on American interests goes beyond Iran. Such cooperation has the potential, basically, to remake the geopolitics of all Eurasia; to establish Moscow as a leading energy supplier, not just to Europe, but also to Asia; to have Moscow as the major influence on energy trade in this part of the world and to consolidate the Sino-Russian axis of oil as the leading counterweight to American hegemony in regional and international affairs.

Now, what can we do about this? I will briefly throw out three ideas.

One is, we have to start taking energy seriously as a foreign policy issue and prioritizing energy security relative to other foreign policy objectives. Let me give you an example. Does it make sense for the United States to push for Ukrainian accession to NATO when domestic politics in the Ukraine do not, in the end, support that objective? It involves further strategic alienation between the United States and Russia. At a time when we're looking for Russian cooperation on energy, on Iran, on a whole host of other issues, does it make sense to go down that road or does it make more sense to incorporate energy interests in a broad strategic dialog with Russia and recognize that on some objectives that we have regarding Russia, that energy security may be more important than those priorities?

The second point is, I think we need a grand bargain between the United States and Iran.

Senator DOMENICI. What did you say that was, sir? Repeat that.

Dr. LEVERETT. I think there needs to be a grand bargain between the United States and the Islamic Republic of Iran. My criticism of the Baker-Hamilton Iraq Study Group recommendations on engaging Iran is not that they go too far, but that they don't go far enough. Unless there is a comprehensive deal between the United States and Iran in which all of the major bilateral differences between the U.S. and Iran are resolved in a package, not only will there be no diplomatic solution to the nuclear issue, but basically, the United States will lose the race for Iran that I described to you a few minutes ago. I think it is very important that the United States embrace a comprehensive wrap approach with Iran as an important foreign policy objective.

Third, I think it is critical that we take seriously the goal of encouraging the internationalization of Chinese and Indian state-owned energy companies. These companies are, despite their state-owned status, in many ways becoming more market-oriented, more profit-focused in their operations, in their strategies, in their planning, and they are increasingly willing, in significant ways, to operate autonomously from their national governments. This is a trend that we ought to be encouraging. In that regard, the political response in this country to the possibility that a Chinese energy company, which last year made a serious initiative to buy Unocal, I think, sent exactly the wrong message to the Chinese. We need to be encouraging the Chinese and the Indians to rely more on the market to meet their energy needs and, instead, in many of our political responses to them, we are sending a message that the United States will not let the market work in ways that will meet their needs, which will only encourage what are, from our perspective, the worst aspects of their current policies. Thank you very much.

[The prepared statement of Dr. Leverett follows:]

PREPARED STATEMENT OF DR. FLYNT LEVERETT*, SENIOR FELLOW AND DIRECTOR, GEOPOLITICS OF ENERGY INITIATIVE, NEW AMERICA FOUNDATION, WASHINGTON, DC

Mr. Chairman, Senator Domenici, members of the Committee, thank you for the opportunity to speak with you about the global oil balance and its implications for America's national security and foreign policy [1]. In my view, the most profound challenges to America's global leadership during the next quarter century are not posed by the risk of strategic failure in Iraq, further proliferation of weapons of mass destruction, or the growth and consolidation of extremist forces in the Islamic world. Rather, the most profound challenges to U.S. preeminence during the next 25 years flow from the strategic and political consequences of ongoing structural shifts in global energy markets, especially the global oil market. Most notably, cooperation between China and Russia on energy matters is bolstering Sino-Russian cooperation on strategic issues, effectively creating a Sino-Russian ''axis of oil'' as the principal counterweight to America's global hegemony.

RESOURCE NATIONALISM AND RESOURCE MERCANTILISM

The basic structural shifts in global energy markets I see boil down to two important trends:

- *The first* is the tightening of margins between global demand for crude oil and installed upstream productive capacity. The global oil supply has grown steadily in recent years, and there is considerable evidence that it will continue to grow for many years to come, given suitable oil prices and appropriate levels of investment. But, in recent years, global demand for crude oil has been growing faster than supply—in no small part because of burgeoning energy demand from emerging economic

* Flynt Leverett is Senior Fellow and Director of the Geopolitics of Energy Initiative at the New America Foundation. He is also a visiting professor of political science at the Massachusetts Institute of Technology and a principal of Strategic Energy and Global Analysis, LLC. Previously, he served in government as senior director for Middle East affairs at the National Security Council, on the Secretary of State's Policy Planning Staff, and as senior analyst at the Central Intelligence Agency.

powerhouses in Asia, particularly China and India. In coming years, demand is likely to continue bumping up against installed productive capacity.

- *The second* important structural shift in the global oil market is the progressive concentration of the world's oil reserves under the control of national governments and national oil companies, especially in the Middle East and the former Soviet Union.

Taken together, these two trends are generating strategic and political responses on both the supply side and the demand side of the global oil market. On the supply side, many have noted the rise of ''resource nationalism''. Resource nationalism is often defined as national governments' assertion of ownership rights over oil and gas reserves against the interests of international energy companies. But there is another dimension to resource nationalism on which I want to focus—that is, national governments making decisions about the production and marketing of the hydrocarbon reserves under their control not only on the basis of economic factors, but also on the basis of strategic and political calculations.

There are many examples of how resource nationalism can challenge a wide range of American interests. These include:

- Russia's application of energy ''levers'' to reestablish its hegemonic position in the post-Soviet space and bolster its strategic position vis-a`-vis Europe and East Asia;
- Venezuela's exploitation of its dominant position as a Western hemisphere energy producer and exporter to weaken America's standing in parts of Latin America; and
- Saudi Arabia using its unique status as the ''swing producer'' for the global oil market to cultivate a deepening strategic relationship with China as a ''hedge'' against precipitous deterioration in the Kingdom's traditional strategic partnership with the United States [2]

On the demand side, we are witnessing an analogous phenomenon, which I describe as ''resource mercantilism''—that is, the reliance of energy importing states on national energy companies to secure access to overseas oil and gas resources on more privileged bases than simple supply contracts. Resource mercantilist states provide various kinds of support to their national oil companies' efforts to acquirehydrocarbon assets abroad and, like resource nationalist states, often seem to base their actions in global energy markets on strategic calculations as well as on commercial and economic considerations.

The outstanding exemplars of resource mercantilism today are, of course, China and India, both of which perceive increasingly acute vulnerabilities to their energy security stemming from their growing reliance on imported hydrocarbons to fill critical portions of their energy mix. And, there are a growing number of examples of how resource mercantilism can work against U.S. interests, although not in the way that many observers initially anticipated. In this regard, while increased demand from China and other rising Asian economies has had a very direct effect on global oil prices, there is little evidence that Chinese and Indian ''equity oil'' deals are keeping or will keep an economically or strategically significant part of the world'soil reserves ''locked up'' and unavailable to international markets.

- Currently, oil produced from Chinese and Indian overseas equity assets represents less than one percent of the oil produced and traded worldwide.
- If the most optimistic projections of Chinese and Indian oil and gas acquisitions abroad prove correct, overseas equity oil production by Chinese and Indian national energy companies might represent roughly 2 percent of total worldwideproduction in 2020.

However, statist approaches in the external energy strategies of rising Asian economies are becoming a serious source of geopolitical tension.

- In East Asia, competition between Beijing and Tokyo over a variety of specific energy deals, a bilateral dispute about sovereignty over possible natural gas reserves in the East China Sea, and jockeying over the ultimate destination of aprojected Russian oil pipeline to Asia have all contributed to the deterioration of Sino-Japanese relations in recent years.
- Even more significantly, China's statist approach to external energy initiatives has become a source of geopolitical tension between China and the United States. In particular, China's search for oil is making it a new competitor to the United States for influence, especially in the Middle East, Central Asia, and Africa. This, in turn, is creating new foreign policy as well as commercial options for energy exporting states at odds with U.S. foreign policy goals, including Iran, Sudan, and Syria [3].

The New Axis of Oil

As separate phenomena, resource nationalism and resource mercantilism are posing increasingly serious challenges to U.S. interests around the world. But the challenge to America's global leadership becomes far more profound when these phenomena intersect, as they do in what I have called a ''new axis of oil'' that is acting as a counterweight to American hegemony on a widening range of issues. The heart of this undeclared but increasingly assertive axis is a growing geopolitical partnership between Russia (a major energy producer) and China (the paradigmatic rising consumer) against what both perceive as excessive U.S. unilateralism in world affairs. Sino-Russian collaboration provides the essential frame for a loose and shifting coalition of energy exporting and energy importing states that acts in specific ways to challenge U.S. leadership in world affairs.

The impact of the new axis of oil on American interests has already been felt inthe largely successful Sino-Russian effort to minimize U.S. influence in Central Asia. Sino-Russian cooperation has been critical to the rise of the Shanghai Cooperation Organization, the world's largest regional security organization (in terms of the populations and territory of participating states) and the only regional security organization in the world in which the United States does not participate. Working through the Shanghai Cooperation Organization, Moscow and Beijing have collaborated over the past three years to cap and then roll back the post-9/11 extension of American influence into Central Asia.

The new axis of oil is also reflected in Sino-Russian cooperation to frustrate a significant segment of U.S. policy objectives regarding the Iranian nuclear issue. Both Russia and China have complicated policy agendas toward the Islamic Republic. To be sure, neither Moscow nor Beijing sees Iranian acquisition of a nuclear weapons capability as a Iran as a desirable turn of events. But both are prepared to tolerate a higher-level of Iranian nuclear development than the present U.S. administration. Moreover, each has other interests that it wants to pursue with Iran.

- For Russia, these interests include exporting civil nuclear technology and conventional military equipment. For China, they include cultivating Iran as an energy supplier.
- And, both Moscow and Beijing have interests in collaborating with Tehran in Central Asia to manage Sunni extremist threats there and minimize U.S. influence. To these ends, Russia and China have now included the Islamic Republic in the Shanghai Cooperation Organization as an observer.

In this context, neither Russia nor China will support multilateral sanctions against Iran that would put these various interests at risk. As a result, there is no prospect of getting the United Nations Security Council to impose sanctions on the Islamic Republic that would be stringent enough to leverage changes in Iranian behavior on the nuclear issue.

Even more significantly, Russia and China see the controversy over Iran's nuclear activities as an important issue on which to ''draw lines'' against what both Moscow and Beijing consider excessive U.S. unilateralism in international affairs. In this regard, Russian and Chinese leaders considered the Iraq war a dangerous precedent and are determined not to see that precedent repeated in Iran. In the end, the United States or others may use military force unilaterally to try to delay Iran's nuclear development, but Moscow and Beijing will use their status as permanent members of the Security Council to ensure that there is no plausible international legitimation for such unilateral action.

The Race for Iran

The geopolitical and geoeconomic stakes at play in Iran go well beyond the nuclear controversy. There is now a broader strategic competition underway between the United States, on the one hand, and Russia and China, on the other, concerning Iran's economic and political role in the Middle East and global energy markets in coming decades. The outcome of this competition hinges in considerable measure on which countries will assume leading roles in helping Iran develop its enormous hydrocarbon resources.

Iran's resource base is truly impressive. If one converts Iran's reserves of natural gas—the second-largest in the world, after Russia's—into barrels of oil equivalent and adds them to Iran's proven reserves of conventional oil—the second-largest in the world, after Saudi Arabia's—Iran's hydrocarbon resources are effectively equal to those of Saudi Arabia and significantly greater than those of Russia [4]. Moreover, Iran's low rates of production of crude oil and natural gas, relative to its reserves base, suggest that the Islamic Republic is perhaps the only major energy-producing state with the resource potential to increase production of both oil and gas by orders of magnitude over the next decade or so.

Iran, however, cannot realize this potential without significant infusions of investment capital and transfers of technology from abroad. Since the mid-1990s, U.S. policy has sought to constrain the development of Iran's hydrocarbon resources by barring U.S. energy companies from doing business there and threatening European companies undertaking projects in Iran with secondary sanctions. These policies, combined with a problematic investment climate in the Islamic Republic, have limited investment flows and transfers of technology into Iran's oil and gas sectors. Recently, Iran's Oil Minister publicly acknowledged this.

Some have suggested that insufficient investment in new productive capacity, along with the combined effects of the depletion of already developed oil and gas fields and the growth in its domestic energy demand has put the Islamic Republic's oil and gas exports into a precipitous decline. But, it would be a mistake to assume that, absent rapprochement with the United States, these trends will continue unchecked and put the Islamic Republic in an increasingly precarious economic and strategic position.

Over the last several months, Iranian officials and energy executives have told me that Iran is developing an alternative strategy for increasing its production and exports of crude oil and natural gas, a strategy that does not rely on substantially improved relations with the United States or the West generally. This strategy hasthree principal elements.

- First, Tehran continues to explore the possibility of energy deals with European energy companies that are willing to do business in the Islamic Republic. While some significant European energy companies are reducing their involvement in Iran, there are still prominent Europe-based international energy companies with upstream investments there that are pursuing additional deals.
- Second, Iran is developing ties to state-owned energy companies in other Islamic countries (i.e., Petronas in Malaysia) and, more importantly, to national energy companies in China and India. Chinese companies, in particular, are making commitments to invest substantial amounts of capital in Iran's oil and gas sectors.
- Third, Iran is exploring possibilities for cooperation with Russia to develop itsenergy production and export capabilities. In particular, Tehran is now willing to "coordinate" the marketing of Iranian gas exports with Moscow to ensure that Iran's emergence as a gas exporter does not work against Russia's economic or strategic interests. In return, Moscow has agreed to provide financial and technical support to help Iran boost its natural gas production [5]. In this context, at the most recent summit meeting of the Shanghai Cooperation Organization, Russian President Vladimir Putin and Iranian President Mahmoud Ahmadinejad announced that their two countries would explore possibilities for cooperating to provide energy exports to Asia.

Privately, Iranian officials and energy executives acknowledge that this approach is not the optimal way to develop their country's hydrocarbon resources. But, as a senior Iranian diplomat put it to me recently, Iran ''cannot wait on the West forever.''

The significance of Russian and Chinese cooperation to develop Iran's hydrocarbon resources goes far beyond its impact on the rate at which the Islamic Republic's oil and gas exports increase or decline or on the extent of Tehran's regional and international isolation. Such cooperation has the potential to help Moscow consolidate a position as the leading player in supplying energy resources to major markets inAsia as well as Europe, with considerable attendant strategic benefits. It also has the potential to consolidate a Sino-Russian axis of oil as the principal counterweight to U.S. hegemony in regional and international affairs.

STRATEGIC CHALLENGES AND POLICY RESPONSES

There are other arenas in which structural shifts in the global oil market and strategic and political responses to those shifts pose serious challenges to America's leadership in international affairs. For example, how major energy exporting states—primarily in the Middle East and Russia—handle their enormous and growing current account surpluses is now as important to the management of global economic imbalances and the future of the dollar as the world's leading reserve currency as the decisions of China and other major Asian economies. Here, too, there is considerable potential for a variant of the axis of oil to develop considerable strategic leverage over the United States.

Of course, the foregoing analysis poses the critical question: ''What is to be done?'' The intellectually and politically facile answer to this question is to advocate ''energy independence'' for the United States. Unfortunately, this is not a serious response to the strategic challenges facing our country. Simply put, there is no economically plausible scenario for a strategically meaningful reduction in the dependence of the United States and its allies on imported hydrocarbons during the next quarter century. Reducing our dependence on domestically produced and imported hydrocarbons has many attractions as a policy goal, but we should have no illusions about how rapidly this can be achieved or how soon it can provide meaningful relief to the strategic challenges I have described.

This means, above all, that we must begin to take energy security seriously as a foreign policy issue and prioritize energy security as a national security objective relative to other foreign policy goals. For example, how important is an abortive drive for Ukraine's accession to NATO to American interests compared to securing Russian cooperation with the United States and its allies on energy supplies, as well as cooperation on the Iranian nuclear issue and other pressing problems? Reasonable and honorable people can come up with different answers to this question and others like it, but to avoid addressing the questions is to avoid the responsibilities of political leadership.

Beyond this general proposition, I would suggest two other concrete policy responses to the strategic challenges growing out of trends in the global oil balance. First, it is critical for the United States to pursue a ''grand bargain'' with the Islamic Republic of Iran—that is, a diplomatic process aimed at resolving all of the outstanding bilateral differences between the United States and Iran as a package. My criticism of proposals for issue-specific or step-by-step engagement with Iran, such as those presented in the Iraq Study Group report, is not that these proposals go too far but, rather, that they do not go far enough. By continuing to reject a grand bargain with Tehran, the Bush administration has not only foreclosed any real chance that Iran will accept meaningful long-term restraints on its nuclear activities; it has also put the United States in a losing position in the longer-term geopolitical and geoeconomic struggle over Iran [6].

Second, it is important to induce the leading resource mercantilist states, China and India, away from statist approaches in their external energy strategies, so as to reduce the chances that they will bolster their strategic commitments to an axis of oil as an international counterweight to the United States. In this regard, it is critically important to bring China and India into the International Energy Agency, the OECD's established ''club'' for major energy importing states. Similarly, it is important to encourage the internationalization of Chinese and Indian national energy companies. There is considerable evidence, especially in the Chinese case, that these companies are becoming more market-oriented and profit-focused in their strategies and operations, and are increasingly willing to challenge their national governments over external energy initiatives that d not make commercial sense. In many ways, these companies are the most promising channels for promoting more market-based approaches to external energy policy in China and India.

The CHAIRMAN. Thank you very much.

Let me ask just a few questions and then I'll defer to Senator Domenici and then we'll take Senators in the order that they arrived, going back and forth between the two sides.

Dr. Birol, let me ask you first. You, I think, alluded to the need for additional transparency in the resource estimates or calculations that exist in some of these areas that are very important for us. My impression is, from your testimony, that there is very little, if any, exploration going on by Saudi Aramco at this time in that area, and I wonder if that is the case and also whether that indicates anything about the likelihood of them developing some of these reserves that they are understood to have. Is it possible also that we are misjudging the rates that we're going to see supplies decline from some of those areas?

Dr. BIROL. Thank you, Mr. Chairman. A reserve transparency is needed both for international oil companies and national oil companies. I would definitely suggest the point that some of the international oil companies recently went through difficult times because they didn't assess their own reserves the right way. But looking at the key resource holders, especially in the major producers in the Middle East, the amount of oil left in those countries, exact knowledge about the amount of oil left in those countries is very important knowledge for all of us, for everybody who is on this planet, as all the numbers show that the bulk of the oil in the future will need to come from those countries.

Now, Saudi Arabia is a key player and will remain so for several years to come and the Saudis have the highest reserves in the world. We do believe that Saudi Arabia has enough oil to meet the growth in global oil demand. However, there are two important points here. One, we would like to be sure, as many people in world—I mean, analysts in the world, how much oil is there and what are the feet-by-feet production levels in Saudi Arabia so that we have an overview of the general picture, which would make everybody feel better and give more confidence to the investor.

The second issue, which is, I think, even as crucial as this one, the growth which will come from Saudi Arabia will not be mainly as a function of their reserves but as a function of their willingness to increase the production capacity.

Saudi Arabia has the reserves, Saudi Arabia has the money to transform these reserves to production, but whether or not in the future Saudi Arabia will increase the production as they did in the past, as much as the world demands from them, or they will leave their oil for the next generations. And Saudi Arabia is differently— they will decide what they are going to do. But it is also the consumers' right to recognize that one day, production from those countries in which we do not have excess, free, extra capital, to go directly into production, may change their policies and this may have serious implications for the consumers.

The structure of the oil market is changing, Mr. Chairman. In the past, the money could have access to many oil deposits in the North Sea, the Gulf of Mexico, but in the future, it will not be the case. Therefore, how much oil will come will be decided by a very few number of national oil companies. And again, market conditions may not be the primary determinant when they are making those decisions. So, from that point of view, there are two major uncertainties: one, whether or not we will have the reserves and the money we'll need in the future, and two, it would be very good to have a more transparency on the reserves in all Middle East countries and the rest of the world.

The CHAIRMAN. Let me ask Dr. Leverett, one of the suggestions you had is that we need to encourage—as I think you stated, encourage the internationalization of Chinese and Indian energy companies. Could you be a little more concrete as to what actions we have—what leverage does our Government have to actually bring that about? And how would you proceed, if you were in a position to do so?

Dr. LEVERETT. It's a very good and important question, Senator Bingaman. You're right, the leverage of the U.S. Government in this area is limited. In the end, if American energy companies decide to pursue joint venture projects upstream, downstream, wherever, with Chinese or Indian state-owned companies, they will ultimately make that decision on commercial grounds. But I think that the political climate and the policy framework that is established can have the effect of encouraging or discouraging this kind of cooperation.

A couple of examples. I already mentioned the political response last year, with regard to the possibility of CNOOC taking over Unocal. I'm certainly not going to make a judgment about the relative merits of CNOOC's offer for Unocal versus Chevron's offer from the standpoint of what was better for Unocal shareholders, but the political response in the United States to just the possibility that CNOOC might take over Unocal, I think, essentially sent the wrong message to the Chinese, which is—and they are already very deeply suspicious about this—the United States encourages China to rely more on the market to meet its energy needs, but the Chinese have very profound doubts that when push comes to shove, we will really let the market, as a matter of policy, work in a disinterested way where their interests are a concern. And the political response to the Sino bid last year, I think, reinforced all of those suspicions on their part.

I would also say, if you look at other countries where international energy companies are based, in many cases, particularly in Europe, these companies, with their governments' encouragement, are actively pursuing possibilities for joint ventures with Chinese and Indian companies. It has become an important part of the strategy for some European energy majors and there is a policy framework in Europe that encourages that.

In this country, I would say the policy framework is, at best, ambivalent on the issue of how desirable it is for American energy companies to be cooperating with Chinese and Indian energy companies and I think it would be in our interests in the long run if we looked on those kinds of possibilities more favorably.

The CHAIRMAN. Thank you very much.

Senator Domenici.

Senator DOMENICI. Thank you very much, Senator Bingaman. I'm sure that two of the witnesses having to sit and wait and get this short time to speak to us is rather boring. But I would like to tell you that, at least from my standpoint and I hope from the rest of the Senators, that we would conclude the same. This kind of hearing is very unusual for us. We don't usually take testimony of this type. We're usually arguing about who is buying what kind of car or what we should do about changing the rules regarding the amount of mileage average cars are going to be getting—all very important things, but what you've told us today is just absolutely startling with reference to the future.

I don't know what we have to do to convince both ourselves and the American people that we must change and do things differently. I just don't know what we have to do besides listening to people like you who are spending your brain power trying to tell us what's going on, that just won't work for too long. And I want to thank you for it and only say that I'm really sorry that we don't have more time. And I commend our chairman for suggesting that we open our year this way. But I still don't think we're going to get it out of one hearing, Mr. Chairman. I don't know if we should pick four or five things that they recommend and say, ''These are things that came out of that, why don't we do something about them?'' I don't know yet, but I'm very, very concerned.

Dr. Leverett, I'll invest a couple of minutes with you and then I'm going to go over to Linda Stuntz. You know, among all the important issues that you see occurring and accruing out there in the world, you came up with discussion of Iran and Russia; is that correct?

Dr. LEVERETT. Yes. There are obviously a lot of specific issues that we could take up under this rhetoric, but I think that the question of the possibilities for Russian and Iranian cooperation on energy matters is an issue that has potentially very, very profound geopolitical and geostrategic implications for the United States. Russia and Iran together control almost half of the world's proven reserves of natural gas. If those two countries are cooperating, coordinating in terms of the way they develop and market their gas exports, they could be potentially twice as influential in the global gas trade as Saudi Arabia is in the global oil trade. And I think that within the last 18 months, Russia and Iran have announced their intention to begin cooperating in this area. There is a high-level Russian/Iranian working group set up to do this. A senior official of Gazprom chairs it on the Russian side, the deputy oil minister of Iran chairs it on the Iranian side, and Russia and Iran are discussing an increasingly wide array of potential energy initiatives, marketing projects and pipeline projects that would increase both Iranian and Russian influence in regional energy markets.

Senator DOMENICI. Well, let me say, from my standpoint, and you can try this on and then if you'd like to comment and critique what I've said, I'll ask a question about the nationalization of oil as an asset around the world. You and I are talking about Russia here and Iran and what they are doing together in their self-interests. Is Russia working as hard as it seems to me to regain its international powers in the region, in the large region that might be called Sino? Is it working as hard as I think and are they doing it with a lot of disregard for the rules because they don't necessarily want to follow them?

Dr. LEVERETT. I think the short answer to your question is yes, they are working very hard to do this. President Putin has made it a cardinal element of his foreign policy to restore Russia to what he would see as its rightful status as a great power in world affairs. But in contrast to previous Russian leaders, he sees the economic aspects of power and indicates that Russia, particularly its energy resources, has the foundation for establishing—re-establishing Russia's role as a great power. I think he and his advisors are working very hard, very strategically to maximize Russia's international standing and influence.

Senator DOMENICI. I want to say to my fellow Senators that about 7, 8, 9 years ago, until about 2 years ago, it was very easy to strike a deal with Russia, the remnants of the Soviet Empire. If we just had money to give to them, they'd give us almost anything.

Dr. LEVERETT. Yes.

Senator DOMENICI. That's gone. They have a new kind of identity. You can hardly talk to them about anything. They're not interested and they act about as affluent as they really are. If you want to deal on their terms, fine. If not, go home. And they act just like they have plenty of money in the bank, and it turns out, from what you've said, they do. And it's not from the deal that you're talking about, they've already put some together, pre-Iranian, and wait until they get it done.

Linda, could I ask you, would you simply turn that map around and show us what's up there so we Senators can see that? Will you put that out for them, too?

Now, Senators, right here in front of us, if there is anything— that's fine, go ahead and put it up where they can see it. If there is anything that has startled me in preparation for this hearing and in today's hearing, it's this chart here, which according to our staff, national oil companies control 3⁄4 of the world's oil reserves. Now, national oil companies, the way we're defining it, kind of gets confusing to some of us. Owned by the state is what national means. So state-owned oil companies has 3⁄4 of the world's reserves. Looking at this chart, you're going to see that the bars on the right reflect the size of the reserves held by the world's oil companies. National oil companies are in red and pink. Despite the fact that some focus on large U.S. corporations as the face of big oil in parentheses, ExxonMobile is the largest non-government-owned company and is at number 14. The relative size of the reserves held signified by the length of the bar is also quite striking.

Well, what I'm trying to tell you is that all this argument here in the United States about our oil and gas companies being so big, and so gigantic that they're taking us over, and so gigantic that we ought to rule them, and so gigantic that we ought to tax them more, well, if you want to come to your senses, just put this chart in front of you and it will tell you what's happening to America— the idea of companies owning oil and/or gas reserves, to provide that to the world in some kind of a capitalistic, free-market approach. Linda, could you tell me—and that's startling me in terms of the future of the world—does that bother you at all, the trend that is moving in the direction that that chart seems to show?

Ms. STUNTZ. It concerns me that we don't know about this, that there is so little awareness of this. National oil companies are not new. They have—Saudi Aramco has been around for a long time. What is different—and I think many of the themes were picked up by Dr. Leverett—is that with the price of oil where it is now and where it's been, they don't need Western capital in the way that they have before.

Also, I think, particularly here in Washington, there has been the notion that they don't have access to technology, if they don't get it from Western companies. That's a bit of a myth also. There are global technology companies. There are fabulous engineers in other places in the world. I'm not going to say maybe it's quite as good, but I think we need to disabuse ourselves of the notion that these folks are somehow dependent on us. They're not, and all of the things you've heard, really, I think, is a striking agreement among the panel that we have to figure out how to deal with them.

They are quite capable of making alliances with others and what that means for us then—I just think we need to get our head around that and figure out what that means for policy. So I appreciate your attention to this and I think it is something that needs to be more broadly understood.

Senator DOMENICI. OK, thank you very much.

The CHAIRMAN. Yes, thank you very much.

Senator Tester is next.

Senator TESTER. Thank you, Mr. Chairman. I've got a question, as long you're warmed up, Linda. One of the first things you talked about was the issue of America's energy independence and how it wasn't achievable. There are a lot of different perspectives when you talk about energy and national security, affordability for regular folks, economic and so on. Is it that? And you talked about Draconian measures, as I recall, but it is that America's energy independence is not that important or it is that we don't have the technology to get it done at this point in time, or is it that, from your perspective, from a technological standpoint, it's simply not obtainable?

Ms. STUNTZ. I would say, from an economic standpoint, it's not achievable. We cannot—we could technologically, if we were willing to pay an unlimited amount for the price of gasoline, if we were willing to waive a lot of environmental requirements, impose speed limits and, I mean, a whole host of things which I think would be unacceptable to us as a public, we could, over some period of time, never import a drop of oil. My point is, that is not going to make us independent of what happens, of all these things that you've been hearing about this morning in global oil markets. It's not going to mean that we don't need to care about whether or not Russia is a reliable supplier to Europe because that matters to us, too. It's a fungible commodity.

But we can't—just one more, quickly. I believe Dr. Hormats was right. The key is vulnerability. We need to make ourselves less vulnerable to those vagaries, even though we can never be completely independent of them.

Senator TESTER. OK, thank you.

Dr. Hormats, you talked about development of alternative sources of energy. Just from your perspective, where do you see it—because there are a lot of different alternative energy potentials out there—where do you see the best opportunity?

Dr. HORMATS. Well, I think there are a number of ways. I'll just use a couple. Cellulosic ethanol, in my judgment, is one of the very interesting alternatives for several reasons. One, it doesn't compete with the food chain.

The CHAIRMAN. I think maybe you could pull that microphone out or turn it on or something.

Dr. HORMATS. Yes, sorry. Thank you.

Thank you, Senator. Cellulosic ethanol, I think, is one area that is very important for a number of reasons. It doesn't compete with the food chain. It is basically using waste material, corn stalks or switch grass or a whole variety of other things. And at this point there is not a large, commercially-viable production facility, but most experts think that with the proper amount of Government support, there will be.

Also, the conversion rate is very efficient compared to things like corn or even sugar, so I think that's one area where the farm community benefits substantially. We reduce our dependence and it's a very efficient way of doing things, and with the right kind of incentives, I think you could do that.

Wind energy, I think, still has a lot of potential in this country as well and there are a couple points that enhance this prospect.

One is, if you take—let me just use one example with relationship to the production tax credit for renewable energy resources like wind or geothermal. Part of the problem is that there is a very narrow window for people to get that. If your project is not completed by the time the window is closed, then you don't get it. And the problem is that it's a very—in some cases, a very short-term window. So there are two or three things, or actually several ways in which—the way the laws are written and the way the incentives are written can be helpful in giving a longer-term time perspective to people who are going to invest in these new sources of energy and this would be one of them.

Part of the problem is that when you get the price going down again, as it is today, people say, we don't really need to do this, the crisis is over, we can relax. The problem is, we really need a longer-term perspective and the way the law—the way the regs are written, the way the PTC is written, you should have a broad enough window to give people who are committing capital for a long period of time the assurance that it's going to be there when their project is complete. Otherwise you get a sort of on-again, off-again process.

So the answer of cellulosic ethanol is one among several, but it is just one that I think happens to have a lot of potential at the moment, if we can develop it further and develop the technology and develop the regulatory and the tax support to do it. We've got the capability, because we've got a lot of switch grass and corn stalks and things, and a lot of States would benefit, not to mention national security.

Senator TESTER. Thank you. Just one last question. I don't know if any of you have had the opportunity to look at energy policy in other countries, and I know we're different than anybody else, but if you have, are there—I mean, I've heard three components here, listed fairly regularly, with the exception of Dr. Leverett's presentation, and that is, deal with domestic production efficiencies and biofuels. Are there any other components to other countries' energy policies that could be of benefit to this country? And it's open to anybody.

Dr. BIROL. The International Energy Agency, we have 26 member governments and the United States is, of course, the key one. All the countries are faced with these twin energy traits: the subject of supply, which we are discussing today, and the second one is the environmental challenge, that we discussed a lot today. And in many cases, they are very much interrelated and many countries do look at—in addition to efficiency increase and the alternative fuel, such as biofuels and the wind and hydro, they also look at the nuclear power from very different angles. Nuclear power again is on the agenda of many countries, even the countries who didn't, in the past, for different reasons, ban nuclear in their system.

The reason here is twofold: One, natural gas prices are very, very high, which makes the electricity production from natural gas very expensive, and the second, as we were discussing with Russia, one of the signals of what Russia gave with the Ukraine-Russia disputes there, security of the gas supply is a key issue and, as my colleagues mentioned here, Russia plus Iran makes about 50 percent of the proven gas reserves, two countries in the headlines of the newspapers for energy.

So, therefore, from the security platform, many countries look at nuclear very closely. In Europe, in Asia, they are both in development in developing countries. But in most of the countries, there is one common denominator—it is the increase of energy efficiency. This is, I think, perhaps the most important one that we have to put the emphasis on, because it is very, very cost-effective and easier, compared to nuclear and the others, to implement and get the support of the public. Thank you.

Senator TESTER. Thank you, panel.

Thank you, Mr. Chairman.

The CHAIRMAN. Thank you very much.

Senator Martinez.

Senator MARTINEZ. Mr. Chairman, thank you very much. I want to really commend you and the ranking member for calling this hearing. I think we always tend to say thank you for calling this important hearing, but some hearings are more important than others and I happen to think this one is tremendously important and timely and I thank the panel for being here today.

I represent the State of Florida and we are vitally interested in developing alternative fuel sources in Florida, particularly, as you mentioned, cellulosic ethanol. We view ourselves as a State that can be a part of the future in that regard.

One of the things that I hear is, in order for us to develop a domestic ethanol production capability, that maintaining the tariff on importing Brazilian ethanol is a good thing for us. And I know, Ms. Stuntz, you suggested that that tariff ought to be removed. Help me with those conflicting currents of encouraging domestic development of the industry while at the same time providing a tremendous challenge for that development by taking off the tariff.

Ms. STUNTZ. Senator, it's a good question. I think the realty is, today, the cost of cellulosic ethanol, according to analyses I've seen, are about five times the cost of corn-based ethanol and potentially even higher than biomass-based ethanol. So it's a significant cost premium today. We've got to do work demonstrations and development to bring that cost down because it clearly is the long-term answer.

Removing the tariff so that low-cost Brazilian ethanol could come in and help us at the time, to give us—to me makes sense as a bridge, frankly, to give us time to do the work that is necessary to bring the cost of that cellulosic ethanol down. Because we're already seeing in the marketplace, as corn-based ethanol is ramped up, corn prices are at about an 11-year high. I was looking for Senator Dorgan, someone who would know better than I do offhand. And you're seeing some pressure, so why not—I don't believe allowing Brazilian or other sugar-based ethanol to come in would significantly slow the work that needs to be done. I think it would assist consumers in the bridge period.

Senator MARTINEZ. Would it also maybe help us in developing a distribution system? Because I know that's one of the areas in which we're also way behind.

Ms. STUNTZ. Absolutely. I think we have got to get to work on— again, as that market expands, we are nearing the limit of how far we can go with tanker transportation, barge transportation of ethanol. I know that Brazil ships ethanol by dedicated ethanol pipelines. I have been beginning to hear—and perhaps some of these in the funding—private sector funding are looking at the possibility of dedicated ethanol pipelines in this country. The technical issues may turn out to be less problematic than economic ones in terms of, ''Do you have a sufficient market to justify and be able to finance that?'' Perhaps Dr. Hormats could say more about that.

But I think you're exactly right. It's all got to come together, and again, doing this in a way that allows that infrastructure to develop would make sense.

Senator MARTINEZ. Dr. Hormats.

Dr. HORMATS. Yes, Linda's point is an interesting and a very good one. I think part of the problem is if there is a gathering system, a distribution system for more conventional sources like corn, for instance. Corn has been marketed as a very commercial and efficient commercial marketing system. For things like switch grass or corn stalks or things like that, there is no efficient gathering system and the distribution process is sort of makeshift. There are lots of little local plants that will be developed, but we need to find a much more efficient way of getting it into the commercial system and distributing it in gas pumps and things like that. Therefore, the more volume there is, the more the incentive exists to develop the kind of things that Linda was talking about, to make a more efficient distribution, gathering and marketing system for these products.

Senator MARTINEZ. Shifting to the geopolitical implications of all of it, Dr. Leverett, I traveled to Georgia recently and saw firsthand not only the influence but the bullying, really, of Russia as it relates to a foreign policy interest merging with the supply of natural gas, particularly to that country. I was also intrigued by your recommendation of a very aggressive engagement with Iran. Some things come to mind as we look to that. I agree with you that in an unfortunate world, that would be a desirable thing. However, there seemed to be some very difficult, long-term goals that the Iranian government seems to be pursuing that might be very difficult for us to overcome. You say in your remarks that we have rejected negotiations with Iran—I'm not sure of the exact language you used, but something along those lines. I'm sure there has not been anything on the table from Iran for us to reject, and I wonder how we would deal with, No. 1, their nuclear ambitions and, No. 2, their stated intent of the destruction of the State of Israel.

Dr. LEVERETT. You've asked exactly the right questions about U.S. Iran policy, precisely because the problems between the United States and Iran are so difficult and there is such an enormous amount of baggage in this relationship on both sides that my argument is the only way this problem is going to be resolved is if it is resolved comprehensively. My own view is that Iran will not agree to strategically meaningful limits on its nuclear activities in the absence of what would, in essence, be a security guarantee from the United States, basically a commitment by the United States not to use force to change the borders or the form of government of the Islamic Republic of Iran. There is no way that an American administration can offer that kind of guarantee unless a range of other problems that we have with Iran—its support for terrorist organizations, its attitude toward the Arab-Israeli conflict and so on—are also dealt with. But there is no way that an Iranian government is going to deal with those issues unless there are things on the table relating to U.S. acknowledgement of an important regional role for Iran, normalization of relations, and lifting of U.S. sanctions. At this point, the issues have become so balled up with one another that you can't resolve any of the outstanding differences between the two countries, I think, without resolving all of them.

In fact, in the spring of 2003, the Iranians offered to negotiate on the kind of comprehensive basis that I am describing and the Bush administration rejected that overture, did not pursue it. The United States and Iran did cooperate very, very extensively after 9/11, with regard to Afghanistan, for a period of almost 2 years, with the Iranians playing very constructively in that process. And again, it was the United States that terminated this dialog. There is much about Iran's behavior and its rhetoric that we rightfully find unacceptable and offensive, but the issue is, how are you going to go about changing that behavior and resolving problems?

My position at this point is that you can't do it unless you're prepared to do it comprehensively. I think that the Iranian leadership, even with President Mahmoud Ahmadinejad in office, would be prepared to engage in a serious process with the United States if they believed it was really aimed at the kind of comprehensive resolution I've described. I think that would help us on the nuclear issue. I think that would help us in terms of the larger war on terror. I think it would help us regionally. And as I tried to indicate today, I think it would be enormously beneficial in terms of long-term energy security.

Senator MARTINEZ. Can I comment on that quickly?

The CHAIRMAN. Sure.

Senator MARTINEZ. It will be very quick.

The CHAIRMAN. Because we need to get some other questions in.

Senator MARTINEZ. This may be right and we may be able to get something if we have a broad dialog with Iran, I don't know. But I want to just go back very briefly to two points. One, we had excellent relations with the Shah. And the Shah, who we put on the throne, by the way, turned around during the middle part of the 1970's and did a lot of things that were quite harmful to America's energy interests. Point one is good relations with the country may be possible, but even the guy we put on the throne was very, very unhelpful in the middle part of the 1970's.

The second point is, even if we were able to achieve these kinds of goals, which I think are noble goals, if we can work out all these things—I'm a little skeptical, but you know what? It's possible. The fact is, that shouldn't deter us one bit from doing the kinds of things we're talking about in this country, to develop new sources of energy, new measures of efficiency to reduce our vulnerability, should this not occur. So it is perhaps important, and I don't deny that maybe it could happen, but we've got a job to do domestically quite apart from that and we shouldn't think that that is an answer to the kinds of broader issues that we have been talking about because the best relations with a country in a given situation can turn very sour, as was the case in the Iran in the 1970's, and then again, after the fall of the Shah as well. So we shouldn't be complacent about these kinds of things.

The CHAIRMAN. If we can go on to Senator Cantwell. We're trying to do 7-minute rounds and I think we can do one for each of the Senators still here, if we try to stick to that. Go ahead.

Senator CANTWELL. Thank you, Mr. Chairman.

I think, Dr. Hormats, that's a great lead-in to the questioning that I'd like to go to. It was kind of brought up by the Chairman earlier about technology.

First, let me just say thank you for all the panelists being here and thank you, Dr. Hormats, for talking about the extension of renewable tax credits for longer horizons. I personally believe as long as we're continuing to use oil at the level we are and only giving 2- or 3-year horizons to the renewables, not only are you only going to have limited development, you basically prevent any kind of the manufacturing from happening in the United States—so the job creation, so thank you, General Wald, for mentioning the efficiency standards for automobiles. I hope next week, in the Commerce Committee, we'll have a bipartisan bill that we can start working on and look at that.

But my question goes to—coming from the Northwest, we kind of look at China as a market and not as a competitor. That's because of Microsoft and Boeing and agriculture, and now the Chinese are drinking lots of Starbucks coffee. So we look at—I think we even looked at President Hu's visit a little differently than the rest of the country did. But why not, given that China, I think, in 2009, will become the largest carbon dioxide emitter? A whole decade before, we'd thought we'd have all this pollution. Why not look at this technology issue, of technology transfer, Dr. Leverett and Dr. Hormats? Even from the Government's perspective of the energy efficiency technology that we have as intellectual property as a way to leverage the relationship with China, beef up the bilateral relationship to a substantive level and use that as at least a starting point to move not only a great economic opportunity for the United States but move China away from Iran?

Dr. LEVERETT. I totally agree with you. I think that we need— I've put in my written testimony a number of elements of what I call the United States-China Energy Partnership, which could involve Japan or it could involve——

Senator CANTWELL. Or in Asia, the United States-Asia energy policy.

Dr. LEVERETT. On many levels, United States-Asia—and there are a variety of reasons. The Chinese are increasingly dependent on imported oil and they are making these arrangements with Russia that you've described and it's bringing them closer to Russia since Russia is a big supplier. And, of course, the Middle East, they are conducting diplomacy, which is not entirely consistent with what we believe to be in our interests.

On the other hand, there are a great many areas where our interests do converge, because they are more and more dependent on imported oil and it is important, I think, to get them to think like a consumer and work with the consumer groups, work with the IEA. There is another energy forum that has been developed at a ministerial issue.

Senator CANTWELL. Get them admitted to the IEA.

Dr. HORMATS. Yes. Part of the problem with the IEA is because you have to be a member of the OEC to be a member of the IEA, but it seems to me you can find some relationship to bring them closer, along with India and others, to work more closely with the IEA. And there is another ministerial form that has been developed, separate from the IEA, with producers and consumers.

But your broader point is absolutely correct. We should look at China because of—as an opportunity, as opposed to a threat. They have environmental problems. We have developed in this country, and are developing now, some very efficient ways of using coal in an environmentally responsible way. Clean coal technology is being developed here. We have great technological capabilities that we're developing in our country. If we work with China, we would have a broader market to sell some of those, which would bring more capital to be available, in the first place, to help to develop them. And a broader partnership would be important across the board for coal, for new technologies, for new ways of developing efficient utilization of existing technologies and the cars that the Chinese are building, a large number of the cars are being developed.

And it's a jobs argument here. This is the part of the problem that I think does not get enough attention. If we really get out in front in energy technology, it can be a great source of job creation in this country, for new types of energy, which we can use here and abroad, and therefore, a partnership at a very high political level.

I had suggested to the State Department before the last summit, President Hu did a lot better in Washington than he did—your Washington than he did in this Washington. And the reason was because of the cooperative relationship that was developed on a commercial basis. That could happen in energy. We need a cabinet-level committee to do it. We had one with Japan in the 1970's, we can have one with the Chinese and this would be of huge benefit to reduce tensions, to avoid the conflicts over Sudan, Iran, Iraq and elsewhere, if we do it right. And give it a high level and have someone senior in the NSC do this, which I think is a good idea, and also in the State Department. The State Department has this idea, I know, because I've given it to them, and others have, but they've really not developed it at a high enough level.

Senator CANTWELL. Dr. Leverett?

Dr. LEVERETT. I think you've really—you've hit on a potential win-win initiative. There is tremendous and growing interest in China now, both in the official community and the business community, about doing things to improve the demand side management part of their energy security equation. But part of it is, in a sense, being behind the curve technologically. And to some degree, there are still— I believe they are still are policy restrictions on exporting some of this technology to China. I think this could be an area where Congress could very constructively take initiatives that would drive policy in a constructive direction because this is part of the energy security equation. The Chinese, themselves, want to pay more attention, just as a brief indicator of that, and also kind of responsive to the questions Senator Tester asked, in terms of what other countries are doing. China's fuel efficiency standards for vehicles marketed in China are today more rigorous than those on the books in the United States.

Ms. STUNTZ. May I just comment very briefly, as maybe a little bit of a dissenting view. I think this is a great goal, but we need to go into this with our eyes open. Every 10 days, China is opening a coal plant with the capacity to serve a town the size of Dallas or San Antonio. Most of those coal plants are not controlled even as well as most of the plants in the United States. They don't even have the base technology that we are putting on right now. A fifth of them are actually characterized as illegal because they haven't been approved by the Federal Government of China. So there is, in part because they can't control all that's going on in the provinces. So I think it is a great aspiration, but we need to be clear about what is actually happening on the ground in China in terms of what's being built. And if we could just raise them up to even the sorts of levels that we're having here and then move beyond that for the future, I think that would be great.

Dr. HORMATS. I agree with that. They are going to do it anyway. We could help them make it more efficient and certainly better from an environmental standpoint.

Ms. STUNTZ. I hope we're not saying they should follow the same route that we have. I would hope that we're saying that we would help them leapfrog that process. I'm well aware, having led a delegation to China in November, and we were all excited because we had energy-efficiency technologies in a good form, but I noticed right next to us, at the same hotel, in a much larger venue, was the mining industry of China. So I know exactly the challenge, but to me, if we could help them leapfrog that, isn't that a great incentive for them to become a better economic interest with the United States and helping us? So instead of going to Iran to look for oil, they know that we can give them energy efficiency. And I think the Chinese do want to be more environmentally sensitive than they are today. They know they are going to get a black eye if they don't approach this issue.

Dr. LEVERETT. There are enormous demonstrations in China, by the Chinese, because of the terrible environmental standards. And the health problem is horrible in China for that reason.

The CHAIRMAN. Please make your comment sort of brief. We need to get on to some other questions.

Dr. BIROL. OK, just one comment. We work very closely with China and India, especially in the last couple of years, and just to let you know, at the last board meeting, we voted Chinese minister and Indian minister participants at the IEM meeting, which is very unusual for us. We just signed if we want to deal up the relationship with China and India in that format. And the second point, very briefly, is that there is already the partnership between China, India, the United States and Japan—the Asia-Pacific Partnership—in the technology area that I wanted to bring to your attention. Thank you.

The CHAIRMAN. All right, thank you very much.

Senator Smith.

Senator SMITH. Thank you, Mr. Chairman.

Thanks to each of our witnesses.

Ms. Stuntz, a question to clarify on the issue of corn-based ethanol and its cost versus cellulosic ethanol. You said that it is more expensive; is that the case because it's inherently more expensive to produce into energy or because we don't have the infrastructure and the gathering of the materials in this country?

Ms. STUNTZ. With the technologies we have now, it is inherently more expensive to produce ethanol from the cellulose of these plant feedstock materials than the technology that we now have to produce it basically from the fermentation of corn.

Senator SMITH. In a country where they had such technology, is it more expensive to produce than corn-based ethanol? In other words, if we get it, if we get the infrastructure, what are the costs comparisons between corn and cellulosic?

Ms. STUNTZ. I believe that there are certain types of plant substances—switch grass and so forth—where the efficiency, the conversion efficiency looks like it will actually be better than it is from corn, but we just don't have the technology yet to do that. And I agree with Dr. Hormats, the infrastructure issue is actually on top of that in terms of what the cost would be. Actually, the ethanol infrastructure is still a bit immature also and could be optimized. Beyond that is an ability to gather all these sources of waste, which sometimes gets ignored. And I think a lot of work needs to be done on that, because it will have to be collected, it will have to be ordered from long distances. So it's a ways in the future, but I think there is a great deal of interest now, both by the private sector, importantly, as well as the Government, and it can happen.

Senator SMITH. I think everyone on the panel is saying it should be pedal to the metal on developing these infrastructures and a way to convert this, because that is one of the building blocks for how we get closer to energy independence.

Dr. HORMATS. Could I just say one thing?

The CHAIRMAN. I think you need to turn on that microphone.

Dr. HORMATS. Sure, good point. Thanks.

One difference is, when you're competing with corn, you're competing with the food stock of the country, and the price goes up. All this other stuff—the switch grass and the cornstalks and a lot of this other— wood chips and things, those are essentially waste materials, so the starting point is much cheaper. The difference also is that there are some commercially-viable plants to do corn ethanol already. We haven't really gotten far enough down the chain to develop those commercially viable plants for this other stuff. Some are being—but they won't be quite ready. They're not ready quite yet, and that's why some greater degree of government effort is needed to help them get from the R&D stage to the development stage, because you can get better efficiencies once you get the volume.

Senator SMITH. And your view is that we don't retard R&D and infrastructure development if we take away the protection against importing——

Dr. HORMATS. I don't think so. I haven't really looked at the economics to that point, but I think that the competition from some of this other stuff, if we're in a country that believes in competition, that could help. But I would say the primary thing is to really get the Government incentives right to provide the capability——

Senator SMITH. And if we do that right, then the protectionism that currently exists——

Dr. HORMATS. Then you don't need it, right.

Senator SMITH. We don't need it. And, moreover, it would allow the industry to develop in this country in a way that doesn't create an artificial sort of price that is uncompetitive—grows to be uncompetitive with what the world market will ultimately be.

Dr. HORMATS. Exactly. And it can't be based just on the movement of prices from month to month because then that creates— this has to be a long-term strategy, to give the companies the incentive, and once you do get the critical mass, then you're in a much better position to resist this—to stand up to competition. And you're absolutely right, at that point, more and more people start using it and the volume enhances the efficiency so you don't have an artificially high price indefinitely.

Senator SMITH. One of the things that I think really needs to be emphasized by this Senate and you all with your voices is that American oil industries are not—they are not controlled by the Federal Government. They are private, for-profit industries. There are competitors in the world. An overwhelming percentage of the competitors to Chevron, Exxon, and all the others, now are nation-state oil industries. I guess one of the questions I really have for you and your insights is, what are these nation-state oil industries that are using energy as a political weapon against their neighbors? What kind of investments are they making for the future in terms of exploration, in terms of infrastructure, as opposed to American oil companies that don't have government control?

Dr. BIROL. They do also make investments and they make investments more or less to nation oil companies and international oil companies and the amount of investment they make is more or less the same in terms of dollars. But there is a small trick here. We just saw this picture, and this picture is a static picture. When we look at the future, the contrast will be even sharper because the oil companies own the reserves, which are declining. This is the point I wanted to say that the national oil companies have young fields and rich fields, so the difference between the international oil companies and national oil companies will change, and in favor of the Nation oil companies because international oil companies are losing reserves and they are not able to have major gains to access the new reserves.

So, in coming to the investments, international oil companies are making a lot of investments in order to maintain the current production capacity, to replace what they have, because it is declining. They have to inject money to slow down the decline. And the national oil companies are making investments to increase the production capacity. So from that point of view, both of them are making investments, but in terms of the production growth, it is coming mainly from the national oil companies. And looking at the future, the picture will change drastically, especially after 5 or 6 years.

Senator SMITH. That brings me to my final point, and then I'll turn it back, Mr. Chairman.

This is not a criticism, Dr. Leverett, but I found your testimony especially chilling in terms of national security implications for our country. It reminded me a little bit that we need to return to Dayton with Russia and others. I'm reminded that in doing that, in asking for energy cooperation, we essentially have to admit to their spheres of influence and tacitly agree to their conduct toward their neighbors.

See, to me, whether Ukraine becomes a member of NATO is not our business. I mean, we get a veto, I suppose. That's the business of Ukrainians. And if they want to be free and they meet the qualifications, our policy is, they get to join. It's shared values. So to say we're not—we're going to say to the Ukrainians, you can't. I don't care—it's up to them. I don't really care if they join or not, but that's up to Ukraine, and I don't want to change the standard just because Russia is going to be offended. I don't know where Russia is going. I mean, Mr. Putin has made it very clear to this Senator that he wants his empire back. It's not going to be communist, but it's sort of a capitalist empire that basically is becoming more and more dictatorial and following its historic pattern.

That really worries me. It would worry me if I were an Estonian or a Latvian or a Pole or a Ukrainian or a Georgian, and I don't think we can accede to that. I don't think we can bargain away those kinds of values that the United States has for the sake of energy security. So I want to say that is a concern.

I think our focus needs to be domestically and then just have a really good military capacity to deal with this. When it comes to Iran, sitting down with them, they made it very clear what they would want from us and that is essentially a military domination of the Middle East. That is a horrifying prospect. If I was an Israeli, I know what that means: I'm gone, I'm exterminated. And I don't think we can accede to that in the name of energy cooperation. So I just wanted to say that. If any of you have a comment on that—I mean, those are the stakes and I don't think we can play in that game.

Dr. LEVERETT. Senator, let me respond just briefly on Russia and Iran. First of all, on Iran, I'm not talking about acceding to Iranian military domination of the Middle East. First of all, frankly, Iran has no capability to project significant levels of conventional military capability much beyond its borders.

Senator SMITH. They'll get them. If they get the energy and we cooperate, they can reload their treasury with petrodollars.

Dr. LEVERETT. They are reloading their treasury whether we cooperate with them or not. The issue for me is, in strategic terms, are we better off trying to have some influence over the way that those petrodollars are accumulated and recycled and invested and used, and how countries that have those kinds of sources of influence use those influences, or are we going to basically, as you say, not play in that game and let other countries play, and therefore, accrue the strategic benefits from that? Those are the trends that I see happening right now, with the emergence of the axis of oil. We basically cannot accomplish our policy objectives toward Iran, because other countries are playing in a game that we've essentially taken ourselves out of, and I don't think that's going to serve American interests in the long run.

On Russia, you're right. In the cold war with the Soviet Union, we accepted the notion of spheres of influence, because basically we made a calculation that managing the nuclear balance and preserving international stability was a more important foreign policy interest of the United States than a quixotic effort to push the Soviets out of their empire before the natural course of events forced them out of their empire.

I would argue we are in a somewhat analogous situation vis-a`vis energy. Russia has this status and leverage as a major energy producer. They are using it to accumulate more and more regional and international influence. Do we want them to use that influence in ways that work against our interests or work in favor of our interests? If we want them to work in ways that support our broader policy objectives, we probably are going to have to accord some attention and some legitimacy to things that they care about. That is the nature of strategy. That is the nature of diplomacy.

The CHAIRMAN. Senator Sessions, you're the clean-up hitter. Go right ahead. We're glad to have you on the committee.

Senator SESSIONS. Thank you. It is a fascinating subject and I'm honored to be with you, and on this committee, and to participate with this panel.

With regard to my friend Gordon Smith's comments, I was at Riga recently and there was sort of an off-the-record discussion over the former Soviet Republic being in NATO and one made the point that it's a question of values. Putin had announced that the greatest disaster of the 20th century was the collapse of the Soviet Union and they thought the best thing that happened in the 20th century was the collapse of the Soviet Union. So I guess I'll say to you, there are more matters than just economics. What might seem most rationale to us, our nations oftentimes act directly contrary to their national economic interests, and how we bring that kind of reasoned analysis to it, I don't know. But I do suggest, Dr. Leverett, that you made some very valuable points and it is true that we'd all be better off if we could accomplish what you suggested there. But it's not easy.

It strikes me now that our national crisis—I think all of you will agree and I'll run down this quickly so we can wrap up—is driven by our dependence, is driven by our need for that fuel that goes into our mobile vehicles. I believe it was—you said that at the beginning—we've got electricity, we've got nuclear and coal in large amounts. So that's not a crisis and natural gas also is a player there.

Ms. Stuntz, you suggested—I believe you stated and I think I recall that the Council on Foreign Relations' report indicated an increased role for nuclear power; is that correct?

Ms. STUNTZ. That's correct. We did not spend a lot of time, and did not have the ability to spend a lot of time on that but it seems to us, to get at our dependence on oil for transportation, we need to do all we can on biofuels, but it may not be enough. And if we could move to electric vehicles which then recharge at night—and we don't think a plug-in hybrid is as far away maybe as some others and there have been some exciting developments recently that we can take advantage of—clean coal and nuclear and fuels like that—and turn it into the transportation field and give us that kind of resiliency in the choices that we have now for electricity.

Senator SESSIONS. And electricity, unfortunately we've gotten in the habit, in recent years, when natural gas prices were low, we were using a lot of natural gas to generate electricity and we could actually—and I think the D.C. busses use natural gas. I mean, natural gas is a possible, a mobile vehicle system. I think you'll agree. I see nods there. And it's also much cleaner, is it not, Dr. Hormats, in terms of global warming gases and carbon?

Dr. HORMATS. Sure, yes.

Senator SMITH. So, to me, that would make a nice move, to reduce the natural gas and electricity production and place it more in meeting this mobility challenge. Of course, you're also suggesting, several of you, that when you increase domestic production—I think is something we need to act on. We're doing it with the Gulf Coast. My area of Alabama is supportive of that and it's good for us. Biofuels, cellulosic ethanol is so exciting to me. We have the potential to grow a lot of switch grass. It's something I've been interested in and Auburn University has done a lot of research on.

But it's that—as you suggested, is it a trend? Is the transformation of that cellulosic, dry material into a fuel that's driving the cost—and we're not quite there yet technologically and that's where we need to invest money, would you agree, Dr. Hormats?

Dr. HORMATS. Absolutely. We have to invest the money in the R&D but also in the development plans, to make it commercially viable. Yes, I totally agree. And it's very doable. The technology is there, it just needs to be expanded and commercialized.

Dr. BIROL. May I just add something on biofuels, why it is very important to develop the biofuels and giving incentives? We should put into perspective that the share of biofuels is today far from being a real alternative to oil products. Today, the share of biofuels in the global oil supply is less than 0.8 percent—not even 1 percent. Just a perspective that there is a big room for improvement, therefore large incentives, a lot of time. It's not a very easy solution.

Senator SMITH. General Wald.

General WALD. I think, first of all, I didn't get a chance to speak much here, but everything that has been talked about today has to happen to make this go away. The interesting thing about Iran that was discussed—and I don't disagree that dialog is a good thing, but the assumption is that Iran is going to potentially cooperate in any way. So that would be nice, but we have to plan that they won't. Like Mahmoud Ahmadinejad, the President of Iran, today has regulated and, via government subsidies, developed in 5 years their entire fleet of automobiles via natural gas in Iran. Now that's a strategic issue to me. And they're going in a direction of their own way on this thing.

So I think, in a sophisticated way, we've been disadvantaged with Russia in our negotiations, but I think countries like Azerbaijan, as far as our relationship and vis-a`-vis Iran and our preparation for a consequence with Iran that will not be in our favor, needs to be addressed today. Now, if we were, today, to do everything that was mentioned today on this panel, it will still take us 15 to 20 years to get there. So we have a window of vulnerability for a decade or more that we have to pay attention.

Then the last thing I'd like to mention on this, on the security aspect, is since 1980, the U.S. Government, through military application, has put about $50 billion to $60 billion a year into the Persian Gulf. That doesn't count the current Iraq war or the 1990 Iraq war. And that's good for our country, for security interests, but the problem is, we're subsidizing world energy. There is nobody else in the world doing this, and really, if you look at how much we're paying per gallon, me, as a U.S. citizen today, for gasoline, you could almost say it's $7 a gallon, based on the fact that we're subsidizing world security on this issue. So I think none of these things are silver bullets. We have to do all of them. And I would appeal to you, as the U.S. leadership and as Senators, to do something comprehensive, across the board, as soon as possible.

Senator SESSIONS. Well, I think that's an excellent statement, General Wald. If you're familiar with Tom Freeman's theory that the more wealth these oil countries have, the worse they behave, both domestically and in foreign policy; do you agree with that? It seems to be true. He even had a little chart, the Freedom Chart, I think, that he put with that.

General WALD. I know Tom and I think his ideas are pretty close. I mean, I'm not necessarily sure you can associate the oil wealth with bad behavior, because we have a lot too, but I do think what it does do is give countries that necessarily don't agree with us the freedom to act like they want to.

Senator SESSIONS. It gives them the ability to increase benefits for that citizen by a small amount and use the extra to invest in military ventures and bad behavior, and it seems to be absolutely happening. I do agree. And I'm part of, with Senators Joe Lieberman and Lindsay Graham and a number of others, a caucus that says we should treat the energy question as a matter of national security, and I think some of the comments you've made here today are real chilling. As Senator Smith said, that drives that home, and this panel has made me more convinced that we do need to see that.

Now, Mr. Chairman, I would just say that means a couple of— means one big thing. I'm a free market conservative. I don't believe in silly policies that drive up the cost to consumers for some vague, theoretical, feel-good thing. I'll just tell you, that's my view of it. However, if it is a matter of national security, then maybe we can justify spending more on this issue of transformation and alternative fuels and those kind of things than we would otherwise. And, in fact, I think we're at that point. I'm prepared to think critically as a member of this committee, Mr. Chairman, on how we can utilize our resources to enhance both our environment and our economic independence.

Dr. HORMATS. Can I just add one quick comment? Our group, the Energy Security Leadership Council, is putting together people like General Wald, financial people, business people, and a number of others to try to look at just that element of security. I think all the members, General Wald and myself and a number of others, Robbie Diamond, who is here, and others stand ready to work with you, because the security element is what makes this different from any other commodity. And it's not a market environment, so you have to come up with solutions where the Government plays a role simply because it is not a free market and is vulnerable to the whims of companies and countries whose interests are very much different from those of the United States.

The CHAIRMAN. OK. Thank you very much, Senator Sessions, and thank you all again for being here.

Let me mention two things. First of all, on this issue of biofuels, we've put out a notice that we're going to have an all-day conference that the committee is sponsoring, on February 1, to look at all the different aspects of the biofuels issue. Many stakeholders are already planning to be here. I just wanted to mention that again.

There are quite a few Senators who came and were not able to stay and ask questions. If they have questions, we've advised them that their questions need to be in by the end of business tomorrow. Then we would ask, if any of them are directed at specific members of the panel, if you could try to respond in a couple of weeks. We would appreciate it very much. Thank you again for being here; it was a very useful hearing.

Senator SESSIONS. Thank you, Mr. Chairman.

[Whereupon, at 12 noon, the hearing was adjourned.]

REFERENCES

[1] For more detailed presentations of the ideas offered in this testimony, see Flynt Leverett and Jeffrey Bader, ''Managing China-U.S. Energy Competition in the Middle East'', The Washington Quarterly (December 2005); Flynt Leverett and Pierre Noel, ''The New Axis of Oil'', The National Interest (Summer 2006); and Flynt Leverett, ''The Race for Iran'', The New York Times, June 20, 2006.

[2] On this point, see also Flynt Leverett, ''Reengaging Riyadh'', in Flynt Leverett, ed., The Road Ahead: Middle East Policy in the Bush Administration's Second Term (Washington, DC: TheBrookings Institution Press, 2005).

[3] Statist strategies for accessing hydrocarbon resources around the world with their inclination toward corruption, provision of ''soft'' loans, and offers of investment and aid in unrelated projects and sectors—also challenge the rules-based international order for tradeand investment in energy that the United States has long championed and, in some cases, weaken leverage the Western governments and international financial institutions can use to promote better governance and transparency in oil-producing countries.

[4] In its December 19, 2005 issues, the Oil and Gas Journal lists Iran's reserves of crude oil roughly 133 billion barrels. Same source lists Canada holding world's second-largest oil reserves, 179 billion barrels, putting Iran in third place. However, reserves estimate for Canada includes 175 billion barrels of reserves in oil sands; this justifies the statement that Iran holds the world's second-largest reserves of conventional oil. When one converts natural gas reserves into barrels of oil equivalent (boe), Saudi Arabia has 302.5 boe in combined reserves of oil and natural gas and Iran has 301.7. By way of comparison, Russia's aggregate hydrocarbon reserves—the world's third-largest—are 198.3 boe. I am grateful to Bijan Khajehpour of Atieh Bahar Consulting for sharing the results of his calculations.

[5] According to both Iranian diplomats and current and former Russian officials, a high-level working group has been set up to oversee bilateral energy cooperation. On the Iranian side, the working group is headed by the Deputy Oil Minister; on the Russian side, it is headed by the chief of Gazprom's international activities.

[6] I have written elsewhere on the content and feasibility of a U.S. Iranian grand bargain; see Flynt Leverett, Dealing With Tehran: Assessing U.S. Diplomatic Options Toward Tehran (New York: The Century Foundation, 2006).

APPENDIX: RESPONSES TO ADDITIONAL QUESTIONS

[The following questions were sent to the witnesses. When the answers are received, they will be retained in the committee files.]

U.S. SENATE, COMMITTEE ON ENERGY AND NATURAL RESOURCES,
Washington, DC, January 17, 2007.

Dr. FATIH BIROL,
Chief Economist, Head of the Economic Analysis Division International Energy Agency, Paris, France.

DEAR DR. BIROL: I would like to take this opportunity to thank you for appearing before the Senate Committee on Energy and Natural Resources on Wednesday, January 10, 2007 to give testimony regarding the global oil balance and its implications for the U.S. economic and national security.

I am enclosing a list of questions which have been submitted for the record. If possible, please respond to these questions by email to Amanda Kelly, amandalkelly.energy senate.gov, by Tuesday, February 6, 2007.

Sincerely,

JEFF BINGAMAN,
Chairman.
[Enclosure.]

QUESTIONS FROM SENATOR DOMENICI FOR DR. BIROL

1. Your written testimony indicates that your ''Alternative Case,'' includes savings from fuel-switching in the power sector. What types of power did you assume we would be switching to?
2. At the hearing, you suggested that the chart showing the relative size of reserves held by the Nation's oil companies as a ''static'' picture, and that a projection of future reserves would show an even more dramatic difference between reserves held by state-and investor-owned oil companies. We would appreciate any further information and data reflecting those projections.

QUESTIONS FROM SENATOR DOMENICI FOR ALL WITNESSES

1. At the hearing, I heard concerns expressed regarding the lack of investment in major oil producing regions, particularly the Middle East, by investor-owned corporations. Even in those areas that have not been closed to private investment, political instability, and a lack of a legal infrastructure for entering into and enforcing contracts seem to be major obstacles. How can the U.S. convince other states that legal and business models that encourage private investment are in their interest as well as ours?
2. Do you have any suggestions for how investor owned companies can position themselves to gain greater access to state-held resources? In the past, these corporations have had expertise that was valuable to the national oil companies. To what extent is this still true, and what can
3. Is it inevitable that investor-owned companies are destined to simply fulfill the role of ''service companies'' to producers of nationally-owned resources'? How does that trend affect U.S. energy security'?
4. The fact that an oil company is nationally-owned does not necessarily exclude the benefits of competition and foreign participation. An example is Norway, How can we promote that business model in the world'?
5. Today we heard testimony that, as a result of the lack of access by investor-owned oil companies to the world's oil ''cheapest'' oil reserves, they are increasingly forced to pursue opportunities in older and more technologically challenging fields. Do you have any information regarding the level of investment on the part of investor-owned

companies that will be required for them to maintain their current levels of production? How does the cost to those companies required to maintain their current level of production compare to the investment required by the world's largest nationally-owned companies?

6. Some analysts have suggested that the trend toward privatization occurred when prices were low, and governments wanted to spread risk. Now nationalization is occurring when prices are high, because governments want to capture the profits. There have been proposals in the U.S. to increase taxes on oil companies, which seems to be a different path to the same goal. Whether the policy is direct nationalization, or appropriation of profits through new taxes, doesn't this increase risk to investors, reducing the industry's ability to make new capital investment?

Questions from Senator Cantwell
Question One: Impact of Growing Chinese Fossil Fuel Demands

The growth of China's energy economy is absolutely astounding. On a recent trip there I learned that they are building roughly 1,000 megawatts of electrical capacity per week, most of it using outdated coal-fired technology. In a single year the Chinese are now putting up the equivalent electricity capacity as the entire Spanish electrical grid.

I also learned that while China only became a net oil importer in 1993, with their accelerating rush towards private vehicles—by 2010 China is expected to have 90 times more cars than they did 1990—Chinese exports in 2030 will match those of the U.S. today.

It is also important to note that new projections predict China will overtake the United States as the world's largest carbon dioxide emitter by 2009, a full decade earlier than previous projections.

I share the deep concern of several of our witnesses that if these trends continue, the results for both global oil demand and carbon emissions will be disastrous.

That's why I believe we must proactively engage and collaborate with China on clean energy alternatives for both our nation's and the world's benefit. I am proud that my home state of Washington already has significant trade ties with China, and we view it as a vast potential export market for our homegrown clean energy technologies.

All Witnesses:

How can the United States constructively engage China as a partner in securing a stable global oil economy?

Senior Chinese officials I met with told they fully recognize the economic, environmental, and security vulnerabilities of their growing fossil fuel dependence, but said they saw few alternatives to meeting their country's incredible demand for energy. How can we help get them the clean energy technologies they want and we want to sell to them? Could Chinese adoption of these technologies create the economies of scale that would drive down the production costs of many renewable energy technologies?

Are there likely to be other benefits of such energy collaboration with the Chinese in terms of other U.S. regional security objectives such as in North Korea, Iran, and the former Soviet Republics? For example, China is the number one importer of oil and gas from Iran and they are bound by energy deals valued at around $120 billion dollars.

Dr. Leverett, what do you think the U.S. can the United States do to counter the Sino-Russian energy axis described in your testimony?

QUESTION TWO: IMPORTANCE OF REDUCING U.S. OIL DEMAND IN TRANSPORTATION SECTOR

It is clear that targeting our transportation system's demand for oil is a key component of improving our nation's energy security. And in fact most future global oil demand is linked to transportation use. If we are going to make any difference in our increasingly dire oil addiction, we need a multi-tiered effort that both increases our use of alternative fuels and reduces our overall demand for fossil fuels.

Success of Vehicle Fuel Economy Standards

According to the 2002 report by the National Academies of Science, gasoline consumption (and oil imports) would be 2.8 million barrels per day higher if CAFE standards had not been imposed in the mid 1970s following the Arab Oil Embargo.

Unfortunately, after doubling the average gas mileage of U.S. vehicles CAFE standards stagnated and have not been increased since 1985.

Next week, I look forward to joining a bipartisan coalition in reintroducing our ''Ten-in-Ten Fuel Economy Act,'' which would increase CAFE standards for all passenger vehicles, including light trucks and SUVs, by 10 miles per gallon over 10 years. This balanced and technically feasible bill would save 25 million barrels of oil per day by 2025, the same amount of oil we currently import from the PersianGulf.

All Witnesses:

- How do you think a reduction in demand of 2.5 million barrels per day would help improve our reliance on foreign oil imports?
- Do you believe that increasing the demand for more efficient vehicles in the largest car market in the world might affect the efficiency of vehicles sold inother countries?
- A recent analysis by an Energy Department analysis found that increasing CAFE standards to a combined fleet average of 35 miles per gallon could save as much as 0.5% of our Gross Domestic Product—roughly $60 billion per year or $200 per capita per year in the United States. Given this tremendous benefit, why do you think it has been so difficult to increase CAFE standards?
- *Mr. Wald,* you testified that increasing transportation efficiency is the single most effective step the U.S. can take to improve its energy security. Can you expand on why you think, from a military and national security perspective, it is critical that we increase the overall efficiency of nation's transportation fleet?
- *Mr. Wald,* I also understand that the Defense Department spends $10.6 billion annually on fuel, or 97 percent of the federal government's use, and almost 2 percent of the entire country's use. And that as much of 70 percent of our military convoys in Iraq, which are increasingly at risk from roadside bombs, are carrying fuel.

Given these facts, how could the military benefits from more efficient vehicle technology? Would you say such technologies would qualitatively increase our nation's military strength?

Breakthrough technologies

- Do any of the witnesses see any groundbreaking or game changing technologies that could dramatically change this bleak future oil demand outlook?

- Plug-in hybrids promise to break the historic wall between. the transportation and electricity sector and provide new and diverse alternative energy sources to displace gasoline and diesel.
- What role will biofuels have in replacing world oil demand?
- Would it be in the national interest to pursue an all out effect in vehicle lightweighting technologies such as the use of composites, or better battery technologies for vehicles?

QUESTION THREE: MYSTERY OF HIGH WORLD OIL PRICES

The projected price of oil in competitive world market has been estimated by the Energy Department to be in the range of S 15 per barrel. And several sources estimate the most efficient OPEC production, in particular Saudi Arabia where the bulk of world oil reserves are, cost as little as $2 to S5 per barrel to produce. Just about four years ago OPEC's target world oil price was between $22 and $28 per barrels. Now we seem content with $50 to 60 per barrel oil, and would be happy if it stayed there for years to come!

All Witnesses:

- What caused this dramatic rebaselining in the world oil price? How come $25 per barrel was okay for everyone a few years ago, now consuming nations would be happy to pay double that.
- What does this target price shift mean in terms of real dollars transferred from U.S. consumers to OPEC and other major oil producers?
- Are there ways we can achieve more stability in the world oil price? Why are there no long-term price contracts in the oil business? It would seem be in the interest of both producing and consuming nations to know in advance how much they would be spending or receiving for a certain amount of oil.
- As you know, the United States manages a strategic petroleum reserve and IEA maintains a shared strategic reserve. Have these proven effective in securing reasonable pricing from OPEC? Could larger reserves help stabilize world oil prices, or at least prevent volatility or future supply shocks?
- What is the value in increasing domestic oil production in environmentally sensitive US oil fields? Isn't this just at best a drop in the bucket in terms of global oil supply?

Question Four: Need for Collaboration among Major Oil Consumers rather Than Military Action

For decades now, ensuring our nation's oil security has been pursued by both competitive economic means and political-military means. It now seems that the net effect of overt military action has proven to be at most of questionable value and may even be counterproductive and contributed to the current destabilization we see in the world oil market.

Several witnesses have described that the geographic distribution of the world's fossil fuel reserves means that OPEC will only gain more leverage in setting future world energy prices. This dynamic, combined with the fact that National Oil Companies control 80 to 90 percent of the world's oil supplies, means to me that maybe it would be wiser to focus on collaborative action among major oil consumers, rather than continue to rely on military action.

All Witnesses:

- What do you think the net effect of the Iraq war on the world oil price?
- What is the prospective effect of a total collapse of the Iraq oil sector?
- Could an organization of major oil consuming countries useful or feasible as a counterweight to OPEC?
- Could IEA form the nucleus of such an organization?
- How can we quickly allow China and other major oil consuming countries to participate fully in and SEA framework?

Mr. Wald:

- In addition to being at odds with a collaborative approach to oil conservation and carbon emissions reduction, do you believe the use of US military power to guarantee oil security feasible in world where OPEC members become nuclear weapons states? How can one or two U.S. carriers project compelling power against a nuclear-armed Iran?

Question Five: Iran's Role in the World Oil Market

The "Oil weapon"

As was mentioned in witness testimony, Iran has threatened to use its oil supply as a weapon to achieve its strategic objectives, including building a nuclear energy, and probably a nuclear weapons program.

- How realistic is this threat and what can we do to counter it?
- I understand that while Iran is a major oil exporter, they have a severe refinery shortage and must import much of their gasoline. Is this a potential leverage point for trade sanctions against Iran?

Iran's oil sector evolution

A recent article by Roger Stern suggests that Iran's oil export capacity might dry up within ten to fifteen years, essentially eliminating a vital government revenue source. This is in contrast to the testimony today which shows the potential of expansion to pre Iran-Iraq war levels of over 6 million barrels per day.

- Do you believe the Stern analysis is credible'? Obviously policy implications toward Iran are quite different if they are really not going to be a major oil exporter within a couple years.

Negotiating with Iran

- Dr Leverett's testimony suggests that the US needs to strike a ''grand bargain'' with Iran. What would be the terms of such a bargain as you envision them?

Questions from Senator Sanders

1. *For Ms. Stuntz*—As I understand it, the National Security Council's energy task force has recommended an automatic 4% yearly increase in CAFE standards. I am interested in what factors were considered in making the decision on what the specific yearly increase should be.
2. *For All Members of the Panel*—Unfortunately, not many of you mentioned climate change in your testimony and it only briefly came up during the hearing. This is a major concern for me because the ongoing

burning of fossil fuels—most obvi ously oil—is leading to global warming. Global warming, in turn, is increasing pressure on natural resources across the globe, and many experts are already warning that this pressure could lead to increased civil unrest in many parts of the world, unrest that could lead to armed conflicts. The ''geopolitics of oil'' should at the very least mention the ''geopolitics of global warming.'' With this in mind, why is there not more emphasis on this important issue in your testimony?

QUESTION FROM MR. SESSIONS

1. Many commentators have criticized the usage of the phrase ''energy independence,'' pointing out that it sets up an unrealistic—if not impossible or undesirable—goal for U.S. energy policy. Given the various priorities that we face when discussing our energy policy, particularly when we consider the issue as a matter of national security, how would each member of the panel articulate the over-arching goal of our policy?

U.S. SENATE, COMMITTEE ON ENERGY AND NATURAL RESOURCES,
Washington, DC, January 17, 2007.

Mr. ROBERT D. HORMATS,
Vice Chairman, Goldman Sachs (International), New York, NY.

DEAR MR. HORMATS: I would like to take this opportunity to thank you for appearing before the Senate Committee on Energy and Natural Resources on Wednesday, January 10, 2007 to give testimony regarding the global oil balance and its implications for the U.S. economic and national security.

I am enclosing a list of questions which have been submitted for the record. If possible, please respond to these questions by email to Amanda Kelly, amandalkelly.energy.senate.gov, by Tuesday, February 6, 2007.

Sincerely,
JEFF BINGAMAN, Chairman.
[Enclosure.]

QUESTIONS FROM SENATOR DOMENICI FOR ALL WITNESSES

1. At the hearing, I heard concerns expressed regarding the lack of investment in major oil producing regions, particularly the Middle East, by investor-owned corporations. Even in those areas that have not been closed to private investment, political instability, and a lack of a legal infrastructure for entering into and enforcing contracts seem to be major obstacles. How can the U.S. convince other states that legal and business models that encourage private investment are in their interest as well as ours?
2. Do you have any suggestions for how investor owned companies can position themselves to gain greater access to state-held resources? In the past, these corporations have had expertise that was valuable to the national oil companies. To what extent is this still true, and what can
3. Is it inevitable that investor-owned companies are destined to simply fulfill the role of ''service companies'' to producers of nationally-owned resources? How does that trend affect U.S. energy security?
4. The fact that an oil company is nationally-owned does not necessarily exclude the benefits of competition and foreign participation. An example is Norway. How can we promote that business model in the world?
5. Today we heard testimony that, as a result of the lack of access by investor-owned oil companies to the world's oil ''cheapest'' oil reserves, they are increasingly forced to pursue opportunities in older and more technologically challenging fields. Do you have any information regarding the level of investment on the part of investor-owned companies that will be required for them to maintain their current levels of production? How does the cost to those companies required to maintain their current level of production compare to the investment required by the world's largest nationally-owned companies?
6. Some analysts have suggested that the trend toward privatization occurred when prices were low, and governments wanted to spread risk. Now nationalization is occurring when prices are high, because governments want to capture the profits. There have been proposals in the U.S. to increase taxes on oil companies, which seems to be a different path to the same goal. Whether the policy is direct nationalization, or appropriation of profits through new taxes, doesn't this increase risk to investors, reducing the industry's ability to make new capital investment?

QUESTIONS FROM SENATOR CANTWELL QUESTION ONE: IMPACT OF GROWING CHINESE FOSSIL FUEL DEMANDS

The growth of China's energy economy is absolutely astounding. On a recent trip there I learned that they are building roughly 1,400 megawatts of electrical capacity per week, most of it using outdated coal-fired technology. in a single year the Chinese are now putting up the equivalent electricity capacity as the entire Spanish electrical grid.

I also learned that while China only became a net oil importer in 1993, with their accelerating rush towards private vehicles—by 2010 China is expected to have 90 times more cars than they did 1990—Chinese exports in 2030 will match those of the U.S. today.

It is also important to note that new projections predict China will overtake the United States as the world's largest carbon dioxide emitter by 2009, a full decade earlier than previous projections.

Share the deep concern of several of our witnesses that if these trends continue, the results for both global oil demand and carbon emissions will be disastrous.

That's why I believe we must proactively engage and collaborate with China on clean energy alternatives for both our nation's and the world's benefit. I am proud that my home state of Washington already has significant trade ties with China, and we view it as a vast potential export market for our homegrown clean energy technologies.

All Witnesses:

How can the United States constructively engage China as a partner in securing a stable global oil economy?

Senior Chinese officials I met with told they fully recognize the economic, environmental, and security vulnerabilities of their growing fossil fuel dependence, but said they saw few alternatives to meeting their country's incredible demand for energy. How can we help get them the clean energy technologies they want and we want to sell to them? Could Chinese adoption of these technologies create the economies of scale that would drive down the production costs of many renewable energy technologies?

Are there likely to be other benefits of such energy collaboration with the Chinese in terms of other U.S. regional security objectives such as in North Korea, Iran, and the former Soviet Republics? For example, China is the number one importer of oil and gas from Iran and they are bound by energy deals valued at around $120 billion dollars.

Dr. Leverett, what do you think the U.S. can the United States do to counter the Sino-Russian energy axis described in your testimony?

Question Two: Importance of Reducing U.S. Oil Demand in Transportation Sector

It is clear that targeting our transportation system's demand for oil is a key component of improving our nation's energy security. And in fact most future global oil demand is linked to transportation use. If we are going to make any difference in our increasingly dire oil addiction, we need a multi-tiered effort that both increases our use of alternative fuels and reduces our overall demand for fossil fuels.

Success of Vehicle Fuel Economy Standards

According to the 2002 report by the National Academies of Science, gasoline consumption (and oil imports) would be 2.8 million barrels per day higher if CAFE standards had not been imposed in the mid 1970s following the Arab Oil Embargo. Unfortunately, after doubling the average gas mileage of U.S. vehicles, CAFE standards stagnated and have not been increased since 1985.

Next week, I look forward to joining a bipartisan coalition in reintroducing our ''Ten-in-Ten Fuel Economy Act,'' which would increase CAFE standards for all passenger vehicles, including light trucks and SUVs, by 10 miles per gallon over 10 years. This balanced and technically feasible bill would save 2.5 million barrels of oil per day by 2025, the same amount of oil we currently import from the Persian Gulf.

All Witnesses:

- How do you think a reduction in demand of 2.5 million bands per day would help improve our reliance on foreign oil imports?
- Do you believe that increasing the demand for more efficient vehicles in the largest car market in the world might affect the efficiency of vehicles sold in other countries?
- A recent analysis by an Energy Department analysis found that increasing CAFE standards to a combined fleet average of 35 miles per gallon could

save as much as 0.5% of our Gross Domestic Product—roughly $60 billion per year or $200 per capita per year in the United States. Given this tremendous benefit, why do you think it has been so difficult to increase CAFE standards?

- *Mr. Wald,* you testified that increasing transportation efficiency is the single most effective step the U.S. can take to improve its energy security. Can you expand on why you think, from a military and national security perspective, it is critical that we increase the overall efficiency of nation's transportation fleet?
- *Mr. Wald,* I also understand that the Defense Department spends $10.6 billion annually on fuel, or 97 percent of the federal government's use, and almost 2 percent of the entire country's use. And that as much of 70 percent of our military convoys in Iraq, which are increasingly at risk from roadside bombs, are carrying fuel.

Given these facts, how could the military benefits from more efficient vehicle technology? Would you say such technologies would qualitatively increase our nation's military strength?

Breakthrough technologies

- Do any of the witnesses see any groundbreaking or game changing technologies that could dramatically change this bleak future oil demand outlook?
- Plug-in hybrids promise to break the historic wall between the transportation and electricity sector and provide new and diverse alternative energy sources to displace gasoline and diesel.
- What role will biofuels have in replacing world oil demand?
- Would it be in the national interest to pursue an all out effect in vehicle lightweighting technologies such as the use of composites, or better battery technologies for vehicles?

Question Three: Mystery of High World Oil Prices

The projected price of oil in competitive world market has been estimated by the Energy Department to be in the range of $15 per barrel. And several sources estimate the most efficient OPEC production, in particular Saudi Arabia where the bulk of world oil reserves are, cost as little as $2 to $5 per barrel to produce. Just about four years ago OPEC's target world oil price was between $22 and $28 per barrels. Now we seem content with $50 to 60 per barrel oil, and would be happy if it stayed there for years to come!

All Witnesses:

- What caused this dramatic rebaselining in the world oil price? How come $25 per ban-el was okay for everyone a few years ago, now consuming nations would be happy to pay double that.
- What does this target price shift mean in terTns of real dollars transferred from U.S. consumers to OPEC and other major oil producers?
- Are there ways we can achieve more stability in the world oil price? Why are there no long-term price contracts in the oil business? It would seem be in the interest of both producing and consuming nations to know in advance how much they would be spending or receiving for a certain amount of oil.
- As you know, the United States manages a strategic petroleum reserve and IEA maintains a shared strategic reserve. Have these proven effective in securing reasonable pricing from OPEC? Could larger reserves help stabilize world oil prices, or at least prevent volatility or future supply shocks?
- What is the value in increasing domestic oil production in environmentally sensitive US oil fields? Isn't this just at best a drop in the bucket in terms of global oil supply?

Question Four: Need for Collaboration among Major Oil Consumers rather Than Military Action

For decades now, ensuring our nation's oil security has been pursued by both competitive economic means and political-military means. It now seems that the net effect of overt military action has proven to be at most of questionable value, and may even be counterproductive and contributed to the current destabilization we see in the world oil market.

Several witnesses have described that the geographic distribution of the world'sfossil fuel reserves means that OPEC will only gain more leverage in setting future world energy prices. This dynamic, combined with the fact that

National Oil Companies control 80 to 90 percent of the world's oil supplies, means to me that maybe it would be wiser to focus on collaborative action among major oil consumers, rather than continue to rely on military action.

All Witnesses:

- What do you think the net effect of the Iraq war on the world oil price?
- What is the prospective effect of a total collapse of the Iraq oil sector?
- Could an organization of major oil consuming countries useful or feasible as a counterweight to OPEC?
- Could IEA form the nucleus of such an organization?
- How can we quickly allow China and other major oil consuming countries toparticipate fully in andMA framework?

Mr. Wald:

- In addition to being at odds with a collaborative approach to oil conservation and carbon emissions reduction, do you believe the use of US military power to guarantee oil security feasible in world where OPEC members become nuclear weapons states? How can one or two U.S. carriers project compellingpower against a nuclear-armed Iran?

QUESTION FIVE: IRAN'S ROLE IN THE WORLD OIL MARKET

The "Oil weapon"

As was mentioned in witness testimony, Iran has threatened to use its oil supply as a weapon to achieve its strategic objectives, including building a nuclear energy, and probably a nuclear weapons program.

- How realistic is this threat and what can we do to counter it?
- I understand that while Iran is a major oil exporter, they have a severe refineryshortage and must import much of their gasoline. Is this a potential leverage point for trade sanctions against -Iran?

Iran's oil sector evolution

A recent article by Roger Stern suggests that Iran's oil export capacity might dry up within ten to fifteen years, essentially eliminating a vital government revenue source. This is in contrast to the testimony today which shows the potential of expansion to pre Iran-Iraq war levels of over 6 million barrels per day.

- Do you believe the Stem analysis is credible? Obviously policy implications toward Iran are quite different if they are really not going to be a major oil exporter within a couple years.

Negotiating with Iran

- Dr Leverett's testimony suggests that the US needs to strike a ''grand bargain'' with Iran. What would be the terms of such a bargain as you envision them?

Questions from Senator Sanders

1. *For Ms. Stunt*—As I understand it, the National Security Council's energy task force has recommended an automatic 4% yearly increase in CAFE standards. I am interested in what factors were considered in making the decision on what the specific yearly increase should be.
2. *For All Members of the Panel*—Unfortunately, not many of you mentioned climate change in your testimony and it only briefly came up during the hearing. This is a major concern for me because the ongoing burning of fossil fuels—most obviously oil—is leading to global warming. Global warming, in turn, is increasing pressure on natural resources across the globe, and many experts are already warning that this pressure could lead to increased civil unrest in many parts of the world, unrest that could lead to armed conflicts. The ''geopolitics of oil'' should at the very least mention the ''geopolitics of global warming.'' With this in mind, why is there not more emphasis on this important issue in your testimony?

Question from Senator Sessions

1. Many commentators have criticized the usage of the phrase ''energy independence,'' pointing out that it sets up an unrealistic—if not impossible or undesirable— goal for U.S. energy policy. Given the

various priorities that we face when discussing our energy policy, particularly when we consider the issue as a matter of national security, how would each member of the panel articulate the over-archinggoal of our policy?

U.S. SENATE, COMMITTEE ON ENERGY AND NATURAL RESOURCES,
Washington, DC, January 17, 2007.

Dr. FLYNT LEVERETT,
Senior Fellow, Director, Geopolitics of Energy Initiative, New America Foundation, Washington, DC.

DEAR DR. LEVERETT: I would like to take this opportunity to thank you for appearing before the Senate Committee on Energy and Natural Resources on Wednesday, January 10, 2007 to give testimony regarding the global oil balance and its implications for the U.S. economic and national security.

I am enclosing a list of questions which have been submitted for the record. If possible, please respond to these questions by email to Amanda Kelly, amandalkelly.energy.senate.gov, by Tuesday, February 6, 2007.

Sincerely,

JEFF BINGAMAN,
Chairman.
[Enclosure.]

QUESTIONS FOR ALL WITNESSES

1. At the hearing, I heard concerns expressed regarding the lack of investment in major oil producing regions, particularly the Middle East, by investor-owned corporations. Even in those areas that have not been closed to private investment, political instability, and a lack of a legal infrastructure for entering into and enforcingcontracts seem to be major obstacles. How can the U.S. convince other states that legal and business models that encourage private investment are in their interest as well as ours?

2. Do you have any suggestions for how investor owned companies can position themselves to gain greater access to state-held resources? In the past, these corporations have had expertise that was valuable to the national oil companies. To what extent is this still true, and what can
3. Is it inevitable that investor-owned companies are destined to simply fulfill the role of ''service companies'' to producers of nationally-owned resources? How does that trend affect U.S. energy security?
4. The fact that an oil company is nationally-owned does not necessarily exclude the benefits of competition and foreign participation. An example is Norway. How can we promote that business model in the world?
5. Today we heard testimony that, as a result of the lack of access by investor-owned oil companies to the world's oil ''cheapest'' oil reserves, they are increasingly forced to pursue opportunities in older and more technologically challenging fields. Do you have any information regarding the level of investment on the part of investor-owned companies that will be required for them to maintain their current levels of production? How does the cost to those companies required to maintain their current level of production compare to the investment required by the world's largest nationally-owned companies?
6. Some analysts have suggested that the trend toward privatization occurred when prices were low, and governments wanted to spread risk. Now nationalization is occurring when prices are high, because governments want to capture the profits. There have been proposals in the U.S. to increase taxes on oil companies, which seems to be a different path to the same goal. Whether the policy is direct nationalization, or appropriation of profits through new taxes, doesn't this increase risk to investors, reducing the industry's ability to make new capital investment?

Question from Senator Cantwell
Question One: Impact of Growing Chinese Fossil Fuel Demands

The growth of China's energy economy is absolutely astounding. On a recent trip there I learned that they are building roughly 1,000 megawatts of electrical capacity per week, most of it using outdated coal-fired technology. in a single year the Chinese are now putting up the equivalent electricity capacity as the entire Spanish electrical grid.

I also learned that while China only became a net oil importer in 1993, with their accelerating rush towards private vehicles—by 2010 China is expected to have 90 times more cars than they did 1990—Chinese exports in 2030 will match those of the U.S. today.

It is also important to note that new projections predict China will overtake the United States as the world's largest carbon dioxide emitter by 2009, a full decade earlier than previous projections.

I share the deep concern of several of our witnesses that if these trends continue, the results for both global oil demand and carbon emissions will be disastrous.

That's why I believe we must proactively engage and collaborate with China on clean energy alternatives for both our nation's and the world's benefit. I am proud that my home state of Washington already has significant trade ties with China, and we view it as a vast potential export market for our homegrown clean energy technologies.

All Witnesses:

- How can the United States constructively engage China as a partner in securing a stable global oil economy?
- Senior Chinese officials I met with told they fully recognize the economic, environmental, and security vulnerabilities of their growing fossil fuel dependence, but said they saw few alternatives to meeting their country's incredible demand for energy. How can we help get them the clean energy technologies they want and we want to sell to them? Could Chinese adoption of these technologies create the economies of scale that would drive down the production costs of many renewable energy technologies?
- Are there likely to be other benefits of such energy collaboration with the Chinese in terms of other U.S. regional security objectives such as in North Korea, Iran, and the former Soviet Republics? For example, China is the number one importer of oil and gas from Iran and they are bound by energy deals valued at around $120 billion dollars.
- Dr. Leverett, what do you think the U.S. can the United States do to counter the Sino-Russian energy axis described in your testimony?

QUESTION TWO: IMPORTANCE OF REDUCING U.S. OIL DEMAND IN TRANSPORTATION SECTOR

It is clear that targeting our transportation system's demand for oil is a key component of improving our nation's energy security. And in fact most future global oil demand is linked to transportation use. If we are going to make any difference in our increasingly dire oil addiction, we need a multi-tiered effort that both increases our use of alternative fuels and reduces our overall demand for fossil fuels.

Success of Vehicle Fuel Economy Standards

According to the 2002 report by the National Academies of Science, gasoline consumption (and oil imports) would be 28 million barrels per day higher if CAFE standards had not been imposed in the mid 1970s following the Arab Oil Embargo. Unfortunately, after doubling the average gas mileage of U.S. vehicles, CAFE standards stagnated and have not been increased since 1985.

Next week, I look forward to joining a bipartisan coalition in reintroducing our ''Ten-in-Ten Fuel Economy Act,'' which would increase CAFE standards for all passenger vehicles, including light trucks and SU-Vs, by 10 miles per gallon over 10 years. This balanced and technically feasible bill would save 2.5 million barrels of oil per day by 2025, the same amount of oil we currently import from the Persian Gulf.

All Witnesses:

- How do you think a reduction in demand of 2.5 million barrels per day would help improve our reliance on foreign oil imports?
- Do you believe that increasing the demand for more efficient vehicles in the largest car market in the world might affect the efficiency of vehicles sold in other countries?
- A recent analysis by an Energy Department analysis found that increasing CAFE standards to a combined fleet average of 35 miles per gallon could save as much as 0.5% of our Gross Domestic Product—roughly $60 billion per year or $200 per capita per year in the United States. Given this tremendous benefit, why do you think it has been so difficult to increase CAFE standards?
- *Mr. Wald,* you testified that increasing transportation efficiency is the single most effective step the U.S. can take to improve its energy

security. Can you expand on why you think, from a military and national security perspective, it is critical that we increase the overall efficiency of nation's transportation fleet?

- *Mr. Wald,* I also understand that the Defense Department spends $10.6 billionannually on fuel, or 97 percent of the federal government's use, and almost 2 percent of the entire country's use. And that as much of 70 percent of our military convoys in Iraq, which are increasingly at risk from roadside bombs, are carrying fuel.

Given these facts, how could the military benefits from more efficient vehicle technology? Would you say such technologies would qualitatively increase our nation'smilitary strength?

Breakthrough technologies

- Do any of the witnesses see any groundbreaking or game changing technologies that could dramatically change this bleak future oil demand outlook?
- Plug-in hybrids promise to break the historic wall between the transportation and electricity sector and provide new and diverse alternative energy sourcesto displace gasoline and diesel.
- What role will biofuels have in replacing world oil demand?
- Would it be in the national interest to pursue an all out effect in vehicle lightweighting technologies such as the use of composites, or better battery technologies for vehicles?

Question Three: Mystery of High World Oil Prices

The projected price of oil in competitive world market has been estimated by the Energy Department to be in the range of $15 per barrel. And several sources estimate the most efficient OPEC production, in particular Saudi Arabia where the bulk of world oil reserves are, cost as little as $2 to $5 per barrel to produce.

Just aboutfour years ago OPEC's target world oil price was between $22 and $28 per barrels. Now we seem content with $50 to 60 per barrel oil, and would be happy if it stayed there for years to come!

All Witnesses:

- What caused this dramatic rebaselining in the world oil price? How come $25 per barrel was okay for everyone a few years ago, now consuming nations wouldbe happy to pay double that.
- What does this target price shift mean in terms of real dollars transferred from U.S. consumers to OPEC and other major oil producers?
- Are there ways we can achieve more stability in the world oil price? Why are there no long-term price contracts in the oil business? It would seem be in the interest of both producing and consuming nations to know in advance how muchthey would be spending or receiving for a certain amount of oil.
- As you know, the United States manages a strategic petroleum reserve and IEA maintains a shared strategic reserve. Have these proven effective in securing reasonable pricing from OPEC? Could larger reserves help stabilize world oil prices, or at least prevent volatility or future supply shocks?
- What is the value in increasing domestic oil production in environmentally sensitive US oil fields? Isn't this just at best a drop in the bucket in terms of global oil supply?

Question Four: Need for Collaboration among Major Oil Consumers rather Than Military Action

For decades now, ensuring our nation's oil security has been pursued by both competitive economic means and political-military means. It now seems that the net effect of overt military action has proven to be at most of questionable value, and may even be counterproductive and contributed to the current destabilization we see in the world oil market.

Several witnesses have described that the geographic distribution of the world's fossil fuel reserves means that OPEC will only gain more leverage in setting future world energy prices.

This dynamic, combined with the fact that National Oil Companies control 80 to 90 percent of the world's oil supplies means to me that maybe it would be wiser to focus on collaborative action among major oil consumers, rather than continue to rely on military action.

All Witnesses:

- What do you think the net effect of the Iraq war on the world oil price?
- What is the prospective effect of a total collapse of the Iraq oil sector?
- Could an organization of major oil consuming countries useful or feasible as a counterweight to OPEC?
- Could IEA form the nucleus of such an organization?
- How can we quickly allow China and other major oil consuming countries to participate fully in and IEA framework?

Mr. Wald:

- In addition to being at odds with a collaborative approach to oil conservation and carbon emissions reduction, do you believe the use of US military power to guarantee oil security feasible in world where OPEC members become nuclear weapons states? How can one or two U.S. carriers project compelling power against a nuclear—-armed Iran?

QUESTION FIVE: IRAN'S ROLE IN THE WORLD OIL MARKET

The ''Oil weapon''

As was mentioned in witness testimony, Iran has threatened to use its oil supply as a weapon to achieve its strategic objectives including building a nuclear energy, and probably a nuclear weapons program.

- How realistic is this threat and what can we do to counter it?
- I understand that while Iran is a major oil exporter, they have a severe refinery shortage arid must import much of their gasoline. Is this a potential leverage point for trade sanctions against Iran?

Iran's oil sector evolution

A recent article by Roger Stern suggests that Iran's oil export capacity might dry up within ten to fifteen years, essentially eliminating a vital government revenue source. This is in contrast to the testimony today which shows the potential of expansion to pre Iran-Traq war levels of over 6 million barrels per day.

- Do you believe the Stern analysis is credible? Obviously policy implications toward Iran are quite different if they are really not going to be a major oil exporter within a couple years.

Negotiating with Iran

- Dr Leverett's testimony suggests that the US needs to strike a ''grand bargain'' with Iran. What would be the terms of such a bargain as you envision them?

QUESTIONS FROM SENATOR SANDERS

1. *For Ms. Stuntz*—As I understand it, the National Security Council's energy task force has recommended an automatic 4% yearly increase in CAFE standards. I am interested in what factors were considered in making the decision on what the specific yearly increase should be.
2. *For All Members of the Panel*—Unfortunately, not many of you mentioned climate change in your testimony and it only briefly came up during the hearing. This is a major concern for me because the ongoing burning of fossil fuels—most obviously oil—is leading to global warming. Global warming, in turn, is increasing pressure on natural resources across the globe, and many experts are already warning that this pressure could lead to increased civil unrest in many parts of the world, unrest that could lead to armed conflicts. The ''geopolitics of oil'' should at the very least mention the ''geopolitics of global warming.'' With this in mind, why is there not more emphasis on this important issue in your testimony?

QUESTION FROM SENATOR SESSIONS

1. Many commentators have criticized the usage of the phrase ''energy independence,'' pointing out that it sets up an unrealistic—if not impossible or undesirable— goal for U.S. energy policy. Given the various priorities that we face when discussing our energy policy, particularly when we consider the issue as a matter of national security,

how would each member of the panel articulate the over-arching goal of our policy?

U.S. SENATE, COMMITTEE ON ENERGY AND NATURAL RESOURCES,
Washington, DC, January 17, 2007.

General CHARLES F. WALD,
USAF (Ret.), Bipartisan Policy Center, Washington, DC.

DEAR GEN. WALD: I would like to take this opportunity to thank you for appearing before the Senate Committee on Energy and Natural Resources on Wednesday, January 10, 2007 to give testimony regarding the global oil balance and its implications for the U.S. economic and national security.

I am enclosing a list of questions which have been submitted for the record. If possible, please respond to these questions by email to Amanda Kelly, amandalkelly.energy.senate.gov, by Tuesday, February 6, 2007.

Sincerely,

JEFF BINGAMAN,
Chairman.

[Enclosure.]

QUESTIONS FROM SENATOR DOMENICI FOR ALL WITNESSES

1. At the hearing, I heard concerns expressed regarding the lack of investment in major oil producing regions, particularly the Middle East, by investor-owned corporations. Even in those areas that have not been closed to private investment, political instability, and a lack of a legal infrastructure for entering into and enforcing contracts seem to be major obstacles. How can the U.S. convince other states that legal and business models that encourage private investment are in their interest as well as ours?
2. Do you have any suggestions for how investor owned companies can position themselves to gain greater access to state-held resources? In the

past, these corporations have had expertise that was valuable to the national oil companies. To what extent is this still true, and what can

3. Is it inevitable that investor-owned companies are destined to simply fulfill the role of ''service companies'' to producers of nationally-owned resources'? How does that trend affect U.S. energy security'?
4. The fact that an oil company is nationally-owned does not necessarily exclude the benefits of competition and foreign participation. An example is Norway, How can we promote that business model in the world'?
5. Today we heard testimony that, as a result of the lack of access by investor-owned oil companies to the world's oil ''cheapest'' oil reserves, they are increasingly forced to pursue opportunities in older and more technologically challenging fields. Do you have any information regarding the level of investment on the part of investor-owned companies that will be required for them to maintain their current levels of production? How does the cost to those companies required to maintain their current level of production compare to the investment required by the world's largest nationally-owned companies?
6. Some analysts have suggested that the trend toward privatization occurred when prices were low, and governments wanted to spread risk. Now nationalization is occurring when prices are high, because governments want to capture the profits. There have been proposals in the U.S. to increase taxes on oil companies, which seems to be a different path to the same goal. Whether the policy is direct nationalization, or appropriation of profits through new taxes, doesn't this increase risk to investors, reducing the industry's ability to make new capital investment?

Questions from Senator Cantwell
Question One: Impact of Growing Chinese Fossil Fuel Demands

The growth of China's energy economy is absolutely astounding. On a recent trip there I learned that they are building roughly 1,000 megawatts of electrical capacity per week, most of it using outdated coal-fired technology. In a single year the Chinese are now putting up the equivalent electricity capacity as the entire Spanish electrical grid.

I also learned that while China only became a net oil importer in 1993, with their accelerating rush towards private vehicles—by 2010 China is expected to have 90 times more cars than they did 1990—Chinese exports in 2030 will match those of the U.S. today.

It is also important to note that new projections predict China will overtake the United States as the world's largest carbon dioxide emitter by 2009, a full decade earlier than previous projections.

I share the deep concern of several of our witnesses that if these trends continue, the results for both global oil demand and carbon emissions will be disastrous.

That's why I believe we must proactively engage and collaborate with China on clean energy alternatives for both our nation's and the world's benefit. I am proud that my home state of Washington already has significant trade ties with China, and we view it as a vast potential export market for our homegrown clean energy technologies.

All Witnesses:

- How can the United States constructively engage China as a partner in securing a stable global oil economy?
- Senior Chinese officials I met with told they fully recognize the economic, environmental, and security vulnerabilities of their growing fossil fuel dependence, but said they saw few alternatives to meeting their country's incredible demand for energy. How can we help get them the clean energy technologies they want and we want to sell to them? Could Chinese adoption of these technologies create the economies of scale that would drive down the production costs of many renewable energy technologies?
- Are there likely to be other benefits of such energy collaboration with the Chinese in terms of other U.S. regional security objectives such as in North Korea, Iran, and the former Soviet Republics? For example, China is the number one importer of oil and gas from Iran and they are bound by energy deals valued at around $120 billion dollars.
- Dr. Leverett, what do you think the U.S. can the United States do to counter the Sino-Russian energy axis described in your testimony?

QUESTION TWO: IMPORTANCE OF REDUCING U.S. OIL DEMAND IN TRANSPORTATION SECTOR

It is clear that targeting our transportation system's demand for oil is a key component of improving our nation's energy security. And in fact most future global oil demand is linked to transportation use. If we are going to make any difference in our increasingly dire oil addiction, we need a multi-tiered effort that both increases our use of alternative fuels and reduces our overall demand for fossil fuels.

Success of Vehicle Fuel Economy Standards

According to the 2002 report by the National Academies of Science, gasoline consumption (and oil imports) would be 2.8 million barrels per day higher if CAFE standards had not been imposed in the mid 1970s following the Arab Oil Embargo. Unfortunately, after doubling the average gas mileage of U.S. vehicles CAFE standards stagnated and have not been increased since 1985.

Next week, I look forward to joining a bipartisan coalition in reintroducing our ''Ten-in-Ten Fuel Economy Act,'' which would increase CAFE standards for all passenger vehicles, including light trucks and SUVs, by 10 miles per gallon over 10 years. This balanced and technically feasible bill would save 25 million barrels of oil per day by 2025, the same amount of oil we currently import from the Persian Gulf.

All Witnesses:

- How do you think a reduction in demand of 2.5 million barrels per day would help improve our reliance on foreign oil imports?
- Do you believe that increasing the demand for more efficient vehicles in the largest car market in the world might affect the efficiency of vehicles sold in other countries?
- A recent analysis by an Energy Department analysis found that increasing CAFE standards to a combined fleet average of 35 miles per gallon could save as much as 0.5% of our Gross Domestic Product—roughly $60 billion per year or $200 per capita per year in the United States. Given this tremendous benefit, why do you think it has been so difficult to increase CAFE standards?
- *Mr. Wald,* you testified that increasing transportation efficiency is the single most effective step the U.S. can take to improve its energy

security. Can you expand on why you think, from a military and national security perspective, it is critical that we increase the overall efficiency of nation's transportation fleet?

- *Mr. Wald,* I also understand that the Defense Department spends $10.6 billion annually on fuel, or 97 percent of the federal government's use, and almost 2 percent of the entire country's use. And that as much of 70 percent of our military convoys in Iraq, which are increasingly at risk from roadside bombs, are carrying fuel.

Given these facts, how could the military benefits from more efficient vehicle technology? Would you say such technologies would qualitatively increase our nation's military strength?

Breakthrough technologies

- Do any of the witnesses see any groundbreaking or game changing technologies that could dramatically change this bleak future oil demand outlook?
- Plug-in hybrids promise to break the historic wall between. the transportation and electricity sector and provide new and diverse alternative energy sources to displace gasoline and diesel.
- What role will biofuels have in replacing world oil demand?
- Would it be in the national interest to pursue an all out effect in vehicle lightweighting technologies such as the use of composites, or better battery technologies for vehicles?

Question Three: Mystery of High World Oil Prices

The projected price of oil in competitive world market has been estimated by the Energy Department to be in the range of S 15 per barrel. And several sources estimate the most efficient OPEC production, in particular Saudi Arabia where the bulk of world oil reserves are, cost as little as $2 to S5 per barrel to produce.

Just about four years ago OPEC's target world oil price was between $22 and $28 per barrels. Now we seem content with $50 to 60 per barrel oil, and would be happy if it stayed there for years to come!

All Witnesses:

- What caused this dramatic rebaselining in the world oil price? How come $25 per barrel was okay for everyone a few years ago, now consuming nations would be happy to pay double that.
- What does this target price shift mean in terms of real dollars transferred from U.S. consumers to OPEC and other major oil producers?
- Are there ways we can achieve more stability in the world oil price? Why are there no long-term price contracts in the oil business? It would seem be in the interest of both producing and consuming nations to know in advance how much they would be spending or receiving for a certain amount of oil.
- As you know, the United States manages a strategic petroleum reserve and IEA maintains a shared strategic reserve. Have these proven effective in securing reasonable pricing from OPEC? Could larger reserves help stabilize world oil prices, or at least prevent volatility or future supply shocks?
- What is the value in increasing domestic oil production in environmentally sensitive US oil fields? Isn't this just at best a drop in the bucket in terms of global oil supply?

Question Four: Need for Collaboration among Major Oil Consumers rather Than Military Action

For decades now, ensuring our nation's oil security has been pursued by both competitive economic means and political-military means. It now seems that the net effect of overt military action has proven to be at most of questionable value and may even be counterproductive and contributed to the current destabilization we see in the world oil market.

Several witnesses have described that the geographic distribution of the world's fossil fuel reserves means that OPEC will only gain more leverage in setting future world energy prices.

This dynamic, combined with the fact that National Oil Companies control 80 to 90 percent of the world's oil supplies, means to me that maybe it would be wiser to focus on collaborative action among major oil consumers, rather than continue to rely on military action.

All Witnesses:

- What do you think the net effect of the Iraq war on the world oil price?
- What is the prospective effect of a total collapse of the Iraq oil sector?
- Could an organization of major oil consuming countries useful or feasible as a counterweight to OPEC?
- Could IEA form the nucleus of such an organization?
- How can we quickly allow China and other major oil consuming countries to participate fully in and SEA framework?

Mr. Wald:

- In addition to being at odds with a collaborative approach to oil conservation and carbon emissions reduction, do you believe the use of US military power to guarantee oil security feasible in world where OPEC members become nuclear weapons states? How can one or two U.S. carriers project compelling power against a nuclear-armed Iran?

Question Five: Iran's Role in the World Oil Market

The "Oil weapon"

As was mentioned in witness testimony, Iran has threatened to use its oil supply as a weapon to achieve its strategic objectives, including building a nuclear energy, and probably a nuclear weapons program.

- How realistic is this threat and what can we do to counter it?
- I understand that while Iran is a major oil exporter, they have a severe refinery shortage and must import much of their gasoline. Is this a potential leverage point for trade sanctions against Iran?

Iran's oil sector evolution

A recent article by Roger Stern suggests that Iran's oil export capacity might dry up within ten to fifteen years, essentially eliminating a vital government revenue source. This is in contrast to the testimony today which shows the potential of expansion to pre Iran-Iraq war levels of over 6 million barrels per day.

- Do you believe the Stern analysis is credible'? Obviously policy implications toward Iran are quite different if they are really not going to be a major oil exporter within a couple years.

Negotiating with Iran

- Dr Leverett's testimony suggests that the US needs to strike a ''grand bargain'' with Iran. What would be the terms of such a bargain as you envision them?

QUESTIONS FROM SENATOR SANDERS

1. *For Ms. Stuntz*—As I understand it, the National Security Council's energy task force has recommended an automatic 4% yearly increase in CAFE standards. I am interested in what factors were considered in making the decision on what the specific yearly increase should be.
2. *For All Members of the Panel*—Unfortunately, not many of you mentioned climate change in your testimony and it only briefly came up during the hearing. This is a major concern for me because the ongoing burning of fossil fuels—most obviously oil—is leading to global warming. Global warming, in turn, is increasing pressure on natural resources across the globe, and many experts are already warning that this pressure could lead to increased civil unrest in many parts of the world, unrest that could lead to armed conflicts. The ''geopolitics of oil'' should at the very least mention the ''geopolitics of global warming.'' With this in mind, why is there not more emphasis on this important issue in your testimony?

QUESTION FROM MR. SESSIONS

1. Many commentators have criticized the usage of the phrase ''energy independence,'' pointing out that it sets up an unrealistic—if not impossible or undesirable— goal for U.S. energy policy. Given the various priorities that we face when discussing our energy policy, particularly when we consider the issue as a matter of national security,

how would each member of the panel articulate the over-arching goal of our policy?

U.S. SENATE, COMMITTEE ON ENERGY AND NATURAL RESOURCES,
Washington, DC, January 17, 2007.

Ms. LINDA G. STUNTZ,
Partner, Stuntz, Davis & Staffier, PC, Washington, DC.

DEAR MS. STUNTZ: I would like to take this opportunity to thank you for appearing before the Senate Committee on Energy and Natural Resources on Wednesday, January 10, 2007 to give testimony regarding the global oil balance and its implications for the U.S. economic and national security.

I am enclosing a list of questions which have been submitted for the record. If possible, please respond to these questions by email to Amanda Kelly, *amandalkelly.energy.senate.gov*, by Tuesday, February 6, 2007.

Sincerely,

JEFF BINGAMAN,
Chairman.

[Enclosure.]

QUESTION FROM SENATOR DOMENICI FOR MS. STUNTZ

1. There has been a great deal of attention paid to the so-called '' '98-'99'' contracts for oil produced in the Gulf of Mexico. These contracts did not contain price ceilings for royalty relief. This was clearly an expensive mistake, but the language is in valid contracts with multiple oil companies, some of which are owned by foreign entities. Some in the other body have proposals that would attempt to alter these contracts, contrary to our contract law. Do we run the risk of setting a bad example for other countries if we don't respect the sanctity of our own government's contracts?

Questions from Senator Domenici for All Witnesses

- At the hearing, I heard concerns expressed regarding the lack of investment in major oil producing regions, particularly the Middle East, by investor-owned corporations. Even in those areas that have not been closed to private investment, political instability, and a lack of a legal infrastructure for entering into and enforcing contracts seem to be major obstacles. How can the U.S. convince other states that legal and business models that encourage private investment are in their interest as well as ours?
- Do you have any suggestions for how investor owned companies can position themselves to gain greater access to state-held resources? In the past, these corporations have had expertise that was valuable to the national oil companies. To what extent is this still true, and what can
- Is it inevitable that investor-owned companies are destined to simply fulfill the role of ''service companies'' to producers of nationally-owned resources'? How does that trend affect U.S. energy security'?
- The fact that an oil company is nationally-owned does not necessarily exclude the benefits of competition and foreign participation. An example is Norway, How can we promote that business model in the world'?
- Today we heard testimony that, as a result of the lack of access by investor-owned oil companies to the world's oil ''cheapest'' oil reserves, they are increasingly forced to pursue opportunities in older and more technologically challenging fields. Do you have any information regarding the level of investment on the part of investor-owned companies that will be required for them to maintain their current levels of production? How does the cost to those companies required to maintain their current level of production compare to the investment required by the world's largest nationally-owned companies?
- Some analysts have suggested that the trend toward privatization occurred when prices were low, and governments wanted to spread risk. Now nationalization is occurring when prices are high, because governments want to capture the profits. There have been proposals in the U.S. to increase taxes on oil companies, which seems to be a different path to the same goal. Whether the policy is direct nationalization, or appropriation of profits through new taxes, doesn't this increase risk to investors, reducing the industry's ability to make new capital investment?

QUESTIONS FROM SENATOR CANTWELL QUESTION ONE: IMPACT OF GROWING CHINESE FOSSIL FUEL DEMANDS

The growth of China's energy economy is absolutely astounding. On a recent trip there I learned that they are building roughly 1,000 megawatts of electrical capacity per week, most of it using outdated coal-fired technology. In a single year the Chinese are now putting up the equivalent electricity capacity as the entire Spanish electrical grid.

I also learned that while China only became a net oil importer in 1993, with their accelerating rush towards private vehicles—by 2010 China is expected to have 90 times more cars than they did 1990—Chinese exports in 2030 will match those of the U.S. today.

It is also important to note that new projections predict China will overtake the United States as the world's largest carbon dioxide emitter by 2009, a full decade earlier than previous projections.

I share the deep concern of several of our witnesses that if these trends continue, the results for both global oil demand and carbon emissions will be disastrous.

That's why I believe we must proactively engage and collaborate with China on clean energy alternatives for both our nation's and the world's benefit. I am proud that my home state of Washington already has significant trade ties with China, and we view it as a vast potential export market for our homegrown clean energy technologies.

All Witnesses:

- How can the United States constructively engage China as a partner in securing a stable global oil economy?
- Senior Chinese officials I met with told they fully recognize the economic, environmental, and security vulnerabilities of their growing fossil fuel dependence, but said they saw few alternatives to meeting their country's incredible demand for energy. How can we help get them the clean energy technologies they want and we want to sell to them? Could Chinese adoption of these technologies create the economies of scale that would drive down the production costs of many renewable energy technologies?

- Are there likely to be other benefits of such energy collaboration with the Chinese in terms of other U.S. regional security objectives such as in North Korea, Iran, and the former Soviet Republics? For example, China is the number one importer of oil and gas from Iran and they are bound by energy deals valued at around $120 billion dollars.
- Dr. Leverett, what do you think the U.S. can the United States do to counter the Sino-Russian energy axis described in your testimony?

QUESTION TWO: IMPORTANCE OF REDUCING U.S. OIL DEMAND IN TRANSPORTATION SECTOR

It is clear that targeting our transportation system's demand for oil is a key component of improving our nation's energy security. And in fact most future global oil demand is linked to transportation use. If we are going to make any difference in our increasingly dire oil addiction, we need a multi-tiered effort that both increases our use of alternative fuels and reduces our overall demand for fossil fuels.

Success of Vehicle Fuel Economy Standards

According to the 2002 report by the National Academies of Science, gasoline consumption (and oil imports) would be 2.8 million barrels per day higher if CAFE standards had not been imposed in the mid 1970s following the Arab Oil Embargo. Unfortunately, after doubling the average gas mileage of U.S. vehicles CAFE standards stagnated and have not been increased since 1985.

Next week, I look forward to joining a bipartisan coalition in reintroducing our ''Ten-in-Ten Fuel Economy Act,'' which would increase CAFE standards for all passenger vehicles, including light trucks and SUVs, by 10 miles per gallon over 10 years. This balanced and technically feasible bill would save 25 million barrels of oil per day by 2025, the same amount of oil we currently import from the Persian Gulf.

All Witnesses:

- How do you think a reduction in demand of 2.5 million barrels per day would help improve our reliance on foreign oil imports?

- Do you believe that increasing the demand for more efficient vehicles in the largest car market in the world might affect the efficiency of vehicles sold in other countries?
- A recent analysis by an Energy Department analysis found that increasing CAFE standards to a combined fleet average of 35 miles per gallon could save as much as 0.5% of our Gross Domestic Product—roughly $60 billion per year or $200 per capita per year in the United States. Given this tremendous benefit, why do you think it has been so difficult to increase CAFE standards?
- *Mr. Wald,* you testified that increasing transportation efficiency is the single most effective step the U.S. can take to improve its energy security. Can you expand on why you think, from a military and national security perspective, it is critical that we increase the overall efficiency of nation's transportation fleet?
- *Mr. Wald,* I also understand that the Defense Department spends $10.6 billion annually on fuel, or 97 percent of the federal government's use, and almost 2 percent of the entire country's use. And that as much of 70 percent of our military convoys in Iraq, which are increasingly at risk from roadside bombs, are carrying fuel.

Given these facts, how could the military benefits from more efficient vehicle technology? Would you say such technologies would qualitatively increase our nation's military strength?

Breakthrough technologies

- Do any of the witnesses see any groundbreaking or game changing technologies that could dramatically change this bleak future oil demand outlook?
- Plug-in hybrids promise to break the historic wall between. the transportation and electricity sector and provide new and diverse alternative energy sources to displace gasoline and diesel.
- What role will biofuels have in replacing world oil demand?
- Would it be in the national interest to pursue an all out effect in vehicle lightweighting technologies such as the use of composites, or better battery technologies for vehicles?

QUESTION THREE: MYSTERY OF HIGH WORLD OIL PRICES

The projected price of oil in competitive world market has been estimated by the Energy Department to be in the range of S 15 per barrel. And several sources estimate the most efficient OPEC production, in particular Saudi Arabia where the bulk of world oil reserves are, cost as little as $2 to S5 per barrel to produce. Just about four years ago OPEC's target world oil price was between $22 and $28 per barrels. Now we seem content with $50 to 60 per barrel oil, and would be happy if it stayed there for years to come!

All Witnesses:

- What caused this dramatic rebaselining in the world oil price? How come $25 per barrel was okay for everyone a few years ago, now consuming nations would be happy to pay double that.
- What does this target price shift mean in terms of real dollars transferred from U.S. consumers to OPEC and other major oil producers?
- Are there ways we can achieve more stability in the world oil price? Why are there no long-term price contracts in the oil business? It would seem be in the interest of both producing and consuming nations to know in advance how much they would be spending or receiving for a certain amount of oil.
- As you know, the United States manages a strategic petroleum reserve and IEA maintains a shared strategic reserve. Have these proven effective in securing reasonable pricing from OPEC? Could larger reserves help stabilize world oil prices, or at least prevent volatility or future supply shocks?
- What is the value in increasing domestic oil production in environmentally sensitive US oil fields? Isn't this just at best a drop in the bucket in terms of global oil supply?

QUESTION FOUR: NEED FOR COLLABORATION AMONG MAJOR OIL CONSUMERS RATHER THAN MILITARY ACTION

For decades now, ensuring our nation's oil security has been pursued by both competitive economic means and political-military means. It now seems that the net effect of overt military action has proven to be at most of questionable value and may even be counterproductive and contributed to the current destabilization we see in the world oil market.

Several witnesses have described that the geographic distribution of the world's fossil fuel reserves means that OPEC will only gain more leverage in setting future world energy prices. This dynamic, combined with the fact that National Oil Companies control 80 to 90 percent of the world's oil supplies, means to me that maybe it would be wiser to focus on collaborative action among major oil consumers, rather than continue to rely on military action.

All Witnesses:

- What do you think the net effect of the Iraq war on the world oil price?
- What is the prospective effect of a total collapse of the Iraq oil sector?
- Could an organization of major oil consuming countries useful or feasible as a counterweight to OPEC?
- Could IEA form the nucleus of such an organization?
- How can we quickly allow China and other major oil consuming countries to participate fully in and SEA framework?

Mr. Wald:

- In addition to being at odds with a collaborative approach to oil conservation and carbon emissions reduction, do you believe the use of US military power to guarantee oil security feasible in world where OPEC members become nuclear weapons states? How can one or two U.S. carriers project compelling power against a nuclear-armed Iran?

Question Five: Iran's Role in the World Oil Market

The ''Oil weapon''

As was mentioned in witness testimony, Iran has threatened to use its oil supply as a weapon to achieve its strategic objectives, including building a nuclear energy, and probably a nuclear weapons program.

- How realistic is this threat and what can we do to counter it?

- I understand that while Iran is a major oil exporter, they have a severe refinery shortage and must import much of their gasoline. Is this a potential leverage point for trade sanctions against Iran?

Iran's oil sector evolution

A recent article by Roger Stern suggests that Iran's oil export capacity might dry up within ten to fifteen years, essentially eliminating a vital government revenue source. This is in contrast to the testimony today which shows the potential of expansion to pre Iran-Iraq war levels of over 6 million barrels per day.

- Do you believe the Stern analysis is credible'? Obviously policy implications toward Iran are quite different if they are really not going to be a major oil exporter within a couple years.

Negotiating with Iran

- Dr Leverett's testimony suggests that the US needs to strike a ''grand bargain'' with Iran. What would be the terms of such a bargain as you envision them?

QUESTIONS FROM SENATOR SANDERS

1. *For Ms. Stuntz*—As I understand it, the National Security Council's energy task force has recommended an automatic 4% yearly increase in CAFE standards. I am interested in what factors were considered in making the decision on what the specific yearly increase should be.
2. *For All Members of the Panel*—Unfortunately, not many of you mentioned climate change in your testimony and it only briefly came up during the hearing. This is a major concern for me because the ongoing burning of fossil fuels—most obviously oil—is leading to global warming. Global warming, in turn, is increasing pressure on natural resources across the globe, and many experts are already warning that this pressure could lead to increased civil unrest in many parts of the world, unrest that could lead to armed conflicts. The ''geopolitics of oil'' should at the very least mention the ''geopolitics of global warming.'' With this in mind, why is there not more emphasis on this important issue in your testimony?

QUESTION FROM MR. SESSIONS

1. Many commentators have criticized the usage of the phrase ''energy independence,'' pointing out that it sets up an unrealistic—if not impossible or undesirable— goal for U.S. energy policy. Given the various priorities that we face when discussing our energy policy, particularly when we consider the issue as a matter of national security, how would each member of the panel articulate the over-arching goal of our policy?

INDEX

B

C

D

G

H

I

J

K

L

M

N

O

P

Q

R